D0162821

An introduction to composite materials

Cambridge Solid State Science Series

DEREK HULL
Professor of Materials Engineering, University of Liverpool

An introduction to composite materials

CAMBRIDGE UNIVERSITY PRESS
Cambridge
London New York New Rochelle
Melbourne Sydney

Published by the Press Syndicate of the University of Cambridge
The Pitt Building, Trumpington Street, Cambridge CB2 1RP
32 East 57th Street, New York, NY 10022, USA
296 Beaconsfield Parade, Middle Park, Melbourne 3206, Australia

First published 1981

Printed in Great Britain at the University Press, Cambridge

British Library cataloguing in publication data

Hull, Derek
An introduction to composite materials.
(Cambridge solid state science series).

1. Composite materials
I. Title
620.1′1892 TA418.9.C6 80-42039

ISBN 0 521 23991 5 hard covers
ISBN 0 521 28392 2 paperback

Contents

Preface

A book on composite materials which is fully comprehensive would embrace large sections of materials science, metallurgy, polymer technology, fracture mechanics, applied mechanics, anisotropic elasticity theory, process engineering and materials engineering. It would have to cover almost all classes of structural materials from naturally occurring solids such as bone and wood to a wide range of new sophisticated engineering materials including metals, ceramics and polymers. Some attempts have been made to provide such an over-view of the subject and there is no doubt that the interaction between different disciplines and different approaches offers a fruitful means of improving our understanding of composite materials and developing new composite systems.

This book takes a rather narrower view of the subject since its main objective is to provide for students and researchers, scientists and engineers alike, a physical understanding of the properties of composite materials as a basis for the improvement of the properties, manufacturing processes and design of products made from these materials. This understanding has evolved from many disciplines and, with certain limitations, is common to all composite materials. Although the emphasis in the book is on the properties of the composite materials as a whole, a knowledge is required of the properties of the individual components: the fibre, the matrix and the interface between the fibre and the matrix.

The essence of composite materials technology is the ability to put strong stiff fibres in the right place, in the right orientation with the right volume fraction. Implicit in this approach is the concept that in making the composite material one is also making the final product. This means that there must be very close collaboration between those who design composite materials at the microscale and those who have to design and manufacture the final engineering component.

Composite materials can be studied at a number of different levels each of which requires a different kind of expertise. The method of approach depends on the objectives of the investigation. Thus, the development of a composite material to resist a corrosive environment, while maintaining its physical and mechanical properties, is primarily an exercise in selecting fibres, resins and interfaces which resist this environment and is within the expertise of chemists, physicists and materials scientists. In contrast, the engineer who has to design a rigid structure, such as an aerodynamic control surface on an aircraft or a pressure pipeline, is more concerned with the macroscopic elastic properties of the material. He uses anisotropic elasticity theory and finite element analysis to design an optimum weight or optimum cost structure with the desired performance characteristics. The disciplines in these two examples barely overlap and yet it is important for the physical scientist to understand the nature of the design problem and for the engineer to appreciate the subtleties of the materials he uses in design. This book goes some way to building the bridge between these widely different approaches and should be of value to all scientists and engineers concerned with composite materials. Naturally, each group will look to other texts for an in-depth treatment of specific aspects of the subject.

The materials scientist has always had a special position in the middle ground between the purely physical and chemical understanding of solids on the one hand and the application of materials in engineering on the other. This book is based on a twenty-lecture course I give to final year honours students reading for BSc or BEng degrees in materials science and materials engineering. In presenting this course I urge all my students to bear in mind an observation by Dr C. C. Chamis on the subject of composite materials, 'The field is still in a fluid state and considerable judgement is required to assess both theoretical and experimental results.' One of the soundest bases for such judgement is a physical understanding of the behaviour of these materials.

The opening chapter of the book deals with the classification and definition of composite materials, the relation between composite materials and more traditional engineering materials and the manufacturing routes for products made from composite materials. In Chapter 2 the properties of fibres and matrices are described with particular reference to the relation between microstructure and property and the influence of processing conditions. Three different classes of fibres, glass, carbon and organic, and two classes of polymer matrices, thermosetting and thermoplastic, are considered.

The interface between the fibre and the matrix and the methods of measuring the bond strength are described in Chapter 3. A brief introduction is given to the physical and chemical principles of adhesion and this is followed by a detailed account, by way of example, of the interface between glass fibres and polyester resin. Other fibre–matrix interfaces are treated in more general terms.

Chapter 4 deals with some geometrical aspects of composite materials with particular reference to the characterisation of fibre volume fraction, fibre length, fibre length distribution, fibre orientation in two and three dimensions, void content, etc. There is also a brief account of the alignment of fibres during elongational and shear flow.

In Chapter 5 the elastic properties of unidirectional laminae, planar-random long fibre laminae and short fibre materials are described. Comparisons are made between theoretical predictions and experimental results for a number of different fibre–resin systems. There is a short section on thermal and curing stresses. The elastic properties of laminates are introduced in Chapter 6 using laminate theory. The mathematical detail is kept to a minimum and special emphasis is given to the physical significance of the analysis. The properties of cross-ply and angle-ply laminates are discussed and the calculation of the stresses in the individual plies is outlined. The results are presented graphically.

The next three chapters are concerned with the strength of laminae, laminates and short fibre materials respectively. Chapter 7 deals with the micromechanisms of failure and the prediction of the strength of unidirectional laminae in longitudinal tension and compression, transverse tension and compression and plane shear. Reference is made to fibre pull-out and fracture energy. The final section deals with the orientation dependence of strength and the use of failure criteria to predict strength in biaxial stress conditions. The strength of multi-ply laminates is described in Chapter 8. The methods used to predict final failure in terms of the progressive failure of the individual laminae are outlined for cross-ply and angle-ply laminates. The effect of free edges on the failure mode and ultimate strength is described. The micromechanisms of fracture and the strength of three groups of short fibre materials are described in Chapter 9. These are short aligned fibres, random in-plane fibres with reference to chopped-strand mat and random short fibres with reference to injection moulded thermoplastic materials.

The final chapter gives an introduction to fatigue and notch sensitivity of composite materials, the deterioration of properties owing to environmental conditions and the development of hybrid materials.

Those who approach composite materials for the first time could easily be overwhelmed by the vast number of scientific and technical papers which have been published in the last few years. To help readers, a list of some of the important publications has been included at the end of each chapter. Students are encouraged to read a selection of these papers but are strongly advised not to attempt to absorb all the detail which they contain. It is far better, at undergraduate level, to obtain a balanced view of the subject.

A book of this kind is possible only because of the pioneering work of many dedicated and distinguished scientists and engineers: Tsai, Kelly, Broutman, Rosen, Watt, Puck to name but a few. I would like to thank many friends and colleagues who have allowed me to use examples from their work and also their publishers who have given me permission to reproduce photographs and diagrams. I am particularly grateful to research students and staff, past and present, in my Department at Liverpool who have contributed in so many ways to my understanding of composite materials. Much of their work has been supported by the Science Research Council and industrial companies which include ICI, BP, Scott–Bader, Pilkington Bros and Ford Motor Company. This support is, and has been, invaluable and is warmly acknowledged.

Derek Hull

1 Introduction

1.1 Classification and definition of composite materials

Composite materials have been classified in many ways depending on the ideas and concepts that need to be identified. A useful and all-embracing classification is set out in Table 1.1 with some examples. Most naturally occurring materials derive their superb properties from a combination of two or more components which can be distinguished readily when examined in optical or electron microscopes. Thus, for example, many tissues in the body, which have high strength combined with enormous flexibility, are made up of stiff fibres such as collagen embedded in a lower stiffness matrix. The fibres are aligned in such a way as to provide maximum stiffness in the direction of high loads and are also able to slide past each other so that the tissue is very flexible. Similarly, a microscopic examination of wood and bamboo reveals a pronounced fibrillar structure which is very apparent in bamboo when it is fractured. It is not surprising that bamboo has been called 'nature's fibre glass'.

Most bulk engineering materials are also combinations of two or more phases dispersed on a microscopic scale to obtain optimum

Table 1.1. *Broad classification of composite materials*

	Examples
Natural composite materials	Wood Bone Bamboo Muscle and other tissue
Microcomposite materials	Metallic alloys: e.g. steels Toughened thermoplastics: e.g. impact polystyrene, ABS Sheet moulding compounds Reinforced thermoplastics
Macrocomposites (Engineering products)	Galvanised steel Reinforced concrete beams Helicopter blades Skis

properties. The strength and toughness of metallic alloys and engineering plastics are achieved by combining high strength phases with tough ductile phases. A relatively simple example is to be found in plain carbon steels (e.g. 99.2% iron 0.8% carbon). When this alloy is cooled slowly from 800 °C the microstructure consists of alternate layers of a soft ductile phase, which is almost pure iron, and a hard brittle compound, Fe_3C, called cementite. Natural materials and engineering materials are both microcomposites since the properties are achieved from a very fine dispersion of the phases. The structure is often so fine that high resolution electron microscopy is required to resolve the separate phases.

The composite 'idea' can be related also to the macroscale. This is particularly relevant to engineering components which may consist of two or more materials combined to give a performance in service which is superior to the properties of the individual materials. Thus, galvanised steel, which is steel coated with a layer of zinc, combines the corrosion resistance of zinc with the strength of steel. Similarly, concrete beams which have excellent compressive strength are given some strength in tension by reinforcing the concrete with steel bars. Helicopter blades combine structural material for strength and stiffness with erosion resistant material to protect the leading edges from damage.

A more relevant classification for the purpose of this book is given in Table 1.2. This is concerned primarily with microcomposite materials and is based on the size, shape and distribution of the two or more phases in the composite. Clearly, the distinction between the different groups is not always a sharp one and the method of making these materials will differ. Most of the metallic alloys achieve their multiphase structure by solid state transformations involving atomic rearrangement and diffusion. Other composite materials can be made by the physical mixing of the separate phases to obtain the desired distribution. The

Table 1.2. *Classification of microcomposite materials*

1. Continuous fibres in matrix: aligned, random
2. Short fibres in matrix: aligned, random
3. Particulates (spheres, plates, ellipsoids, irregular, hollow or solid) in matrix
4. Dispersion strengthened, as for 3 above, with particle size $< 10^{-8}$ m
5. Lamellar structures
6. Skeletal or interpenetrating networks
7. Multicomponent, fibres, particles etc.

composite materials described in this book are in this second category and fall most naturally into the first two groups in Table 1.2.

There is no really adequate definition of a composite material but the preceding discussion gives some indication of the scope of the subject. In terms of the approach to be adopted here there are three main points to be included in a definition of an acceptable composite material for use in structural applications.

(i) It consists of two or more physically distinct and mechanically separable materials.

(ii) It can be made by mixing the separate materials in such a way that the dispersion of one material in the other can be done in a controlled way to achieve optimum properties.

(iii) The properties are superior, and possibly unique in some specific respects, to the properties of the individual components.

The last point provides the main impetus for the development of composite materials. In fibre reinforced plastics, fibres and plastics with some excellent physical and mechanical properties, are combined to give a material with new and superior properties. Fibres have very high strength and modulus but this is only developed in very fine fibres, with diameters in the range 7–15 μm, and they are usually very brittle. Plastics may be ductile or brittle but they usually have considerable resistance to chemical environments. By combining fibres and resin a bulk material is produced with a strength and stiffness close to that of the fibres and with the chemical resistance of the plastic. In addition, it is possible to achieve some resistance to crack propagation and an ability to absorb energy during deformation. This synergistic feature, that is an effect exceeding the sum of the individual effects, of composite materials and structures is best illustrated by the bimetallic strip. This consists of two strips of metal with different thermal coefficients of expansion welded together along their length. Separated, each of these strips remains flat when heated but when they are welded together the composite strip bends towards the metal with the lower coefficient of expansion. The bending is a new and distinct property of the composite strip. The interaction or coupling between materials with different properties is a recurring theme in this book and is the basis for the physical understanding of many of the properties of composite materials.

1.2 Composite materials compared with conventional materials

There has been a rapid growth in the use of fibre reinforced materials in engineering applications in the last few years and there is every indication that this will continue. Some idea of the range of applications for composite materials is given in Table 1.3. The rapid growth has been achieved mainly by the replacement of traditional materials, primarily metals. This suggests that, in some respects, composite materials have superior properties. A comparison between the properties of a range of high strength engineering materials is given in Table 1.4. On the basis of strength and stiffness alone, fibre reinforced composite materials do not have a clear advantage particularly when it is noted that their elongation to fracture is much lower than metals with comparable strength. The advantages of composite materials appear when the modulus per unit weight (specific modulus) and strength per unit weight (specific strength) are considered. The higher specific modulus and specific strength of composite materials means that the weight of components can be reduced. This is a factor of great importance in moving components especially in all forms of transport where reductions in weight result in greater efficiency and energy savings.

In Table 1.4 the properties of aligned composite materials are given both parallel and at right angles to the fibre direction. The very large difference in properties in different directions may be a serious limitation in some applications since the material will be highly anisotropic. However, it is also a source of one of the outstanding advantages of composite materials since it allows the possibility of introducing

Table 1.3. *Applications for fibre reinforced composite materials based on plastics*

Industry	Examples
Aircraft	Wings, fuselages, landing gear, helicopter blades
Automobile	Body parts, lamp-housings, front-end panels, bumpers, leaf springs, seat housings, drive shafts
Boat	Hulls, decks, masts
Chemical	Pipes, tanks, pressure vessels
Furniture and equipment	Panels, housings, chairs, tables, ladders
Electrical	Panels, switchgear, insulators
Sport	Fishing rods, golf clubs, swimming pools, skis, canoes

Table 1.4. *Comparison of some typical values of the properties of engineering materials at 20 °C*

Material	Density ($Mg\ m^{-3}$)	Young's modulus ($GN\ m^{-2}$)	Tensile strength ($MN\ m^{-2}$)	Elongation to fracture (%)	Coefficient of thermal expansion ($10^{-6}\ °C^{-1}$)	Specific Young's modulus, Y. modulus / density ($GN\ m^{-2}$)	Specific tensile strength, T. strength / density ($MN\ m^{-2}$)	Heat resistance (°C)
High strength Al–Zn–Mg alloy	2.80	72	503	11	24	25.7	180	350
Quenched and tempered low alloy steel	7.85	207	2050–600	12–28	11	26.4	261–76	800
Nimonic 90 (nickel-based alloy)	8.18	204	1200	26	16	24.9	147	1100
Nylon 6.6	1.14	2	70	60	90	1.8	61	150
Glass-filled nylon ($V_f = 0.25$)	1.47	14	207	2.2	25	9.5	141	170
Carbon fibre–epoxy resin unidirectional laminae ($V_f = 0.60$)								
(i) parallel to fibres	1.62	220	1400	0.8	−0.2	135	865	260
(ii) perpendicular to fibres	1.62	7	38	0.6	30			
Glass fibre–polyester resin unidirectional laminae ($V_f = 0.50$)								
(i) parallel to fibres	1.93	38	750	1.8	11	19.7	390	250
(ii) perpendicular to fibres	1.93	10	22	0.2				
Glass fibre–polyester resin planar random fibres ($V_f = 0.20$)	1.55	8.5	110	2	25	5.5	71	230

Note: V_f is the volume fraction of fibres.

Table 1.5. *Manufacturing routes for fibre reinforced plastic products*

Manufacturing route	Outline of fabrication and processing methods
Open mould processes	
1. Hand lay-up	Chopped-strand mats, woven roving and other fabrics made from the fibres are placed on the mould and impregnated with resin by painting and rolling. Layers are built up until design thickness is achieved. Moulding cures without heat or pressure.
2. Spray-up	Chopped rovings and resin are sprayed simultaneously into a prepared mould and rolled before the resin cures.
3. Vacuum bag, pressure bag, autoclave	Layers of fibres, usually unidirectional sheets, are pre-impregnated with resin and partially cured (β-staged) to form a pre-preg. The pre-preg sheets are stacked on the mould surfaces in predetermined orientations, covered with a flexible bag, and consolidated using a vacuum or pressure bag in an autoclave at the required curing temperature.
4. Filament winding	Continuous rovings or strands of fibres are fed over rollers and guides through a bath of resin and then wound, using a programme controlled machine, onto a mandrel at pre-determined angles. The resin is partially or completely cured before removing the component, usually a tube, from the mandrel.
5. Centrifugal casting	Mixtures of the fibres and resin are introduced into a rotating mould and allowed to cure *in situ*.
Closed mould processes	
6. Hot press moulding, compression moulding	Heated matched dies or tools are loaded with raw material (sheet moulding compound, SMC, dough moulding compound, DMC, cloth or unidirectional pre-preg) pressed to the shape of the cavity and cured.
7. Injection moulding, transfer moulding	Molten or plasticised polymer mixed with short fibres is injected, usually at high pressure, into the cavity of a split mould and allowed to solidify or cure.
8. Pultrusion	A continuous feed of fibres, in pre-selected orientations, is impregnated with resin and pulled through a heated die to give the shape of the final section (e.g. tubes or I-beams). Partial or complete cure occurs during passage through the die.

Table 1.5 cont.

Manufacturing route	Outline of fabrication and processing methods
9. Cold press moulding	A low pressure, low temperature process in which fibres are impregnated with resin and then pressed between matched dies. Heat is generated during the cure.
10. Resin injection	Fibres in cloth form are placed in the tool which is then closed. The resin is injected at low pressure into the cavity and flows through the fibres to fill the mould space.
11. Reinforced reaction injection moulding (RRIM)	A rapid curing resin system involving two components which are mixed immediately before injection is used. Fibres are either placed in the closed mould before resin is injected or added as short chopped fibres to one of the resin components to form a slurry before injection.

stiffness and strength into a product where it is really required. In other words, it introduces an element of flexibility into design, but design is correspondingly more difficult and demanding.

There are many manufacturing routes for fibre reinforced plastics. Although this book is not concerned primarily with this aspect of the subject, it is important to recognise the profound effect that the manufacturing route and processes have on the final properties of composite materials owing to their effect on the microstructure and internal stresses. A summary of the main manufacturing routes is given in Table 1.5. Where appropriate, brief reference is made to these processes in the text.

1.3 Some composite materials problems: principles

As mentioned in the preface, this book is directed primarily to a physical understanding of the properties of fibre reinforced plastics. It is for this reason that only a limited number of fibres and matrix systems have been considered. Potentially, any material can be used as a matrix for fibres and the scope in terms of properties is almost unbounded. In practice, only a limited number of materials are used and selection is determined by factors such as ease of fabrication, compatibility with fibres, desired end properties and cost. It is not necessary or practical to consider all possible polymer matrices. Thus, only two groups of thermosetting resins, namely polyesters and epoxides, and three thermoplastics, namely nylon 6.6, polycarbonate and polypropylene, are described. The data given in Sections 2.5 and 2.6 are used as a basis for the prediction of the properties of composite materials using these polymers as matrices. Although principles have priority over the absolute values of the results it must be understood that each resin and each fibre has unique properties and it is necessary always to consider each system in its own right.

Similarly, for simplicity, only three classes of fibre are described, namely glass, carbon (Types I and II) and polyamide. Even within this small group there is a very wide range of properties. It is intended to demonstrate how fibre properties, which often depend on the manufacturing route and processing conditions, affect the properties of composite materials. There are many other fibres including boron, high density polyethylene, and a wide range of filamentary materials available.

With an understanding of the principles underlying the behaviour of composite materials, it is possible to approach some of the challenging problems which influence the development of these materials. These problems include, (i) the creep of fibre reinforced thermoplastics, (ii) the reversible and irreversible changes in property which occur owing to contact with humid environments and to temperature fluctuations, (iii) the design of products with optimum fibre content, (iv) the design of composites with energy absorbing capability, (v) the development of materials with resistance to stress and strain corrosion, and (vi) the improvement of the wear resistance of particulate composite materials.

2 Fibres and matrices

2.1 Carbon fibres

High strength, high modulus carbon fibres are about 7 to 8 μm in diameter and consist of small crystallites of 'turbostratic' graphite, one of the allotropic forms of carbon. In a graphite single crystal the carbon atoms are arranged in hexagonal arrays, as illustrated in Fig. 2.1, which are stacked on top of each other in a regular ABAB... sequence. The atoms in the layer or basal planes are held together by very strong covalent bonds and there are weak van der Waal forces between the layers. This means that the basic crystal units are highly anisotropic; the in-plane Young's modulus parallel to the *a*-axes is 910 GN m^{-2} and Young's modulus parallel to the *c*-axis normal to the basal planes is 30 GN m^{-2}. The spacing between the layers is 0.335 nm. Turbostratic graphite closely resembles graphite single crystals except that the layer planes have no regular packing in the *c*-axis direction and the average spacing between the layer planes is about 0.34 nm.

To obtain high modulus and strength the layer planes of the graphite

Fig. 2.1. Arrangement of carbon atoms in the layer planes of graphite.

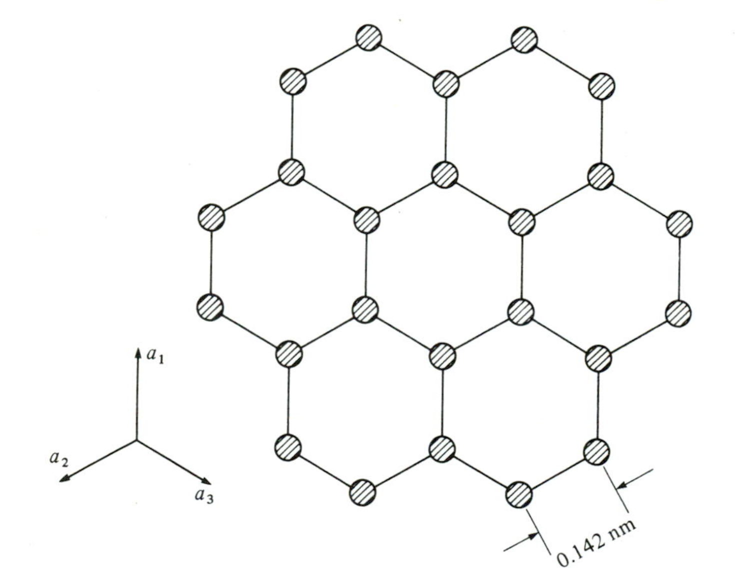

have to be aligned parallel to the axis of the fibre. In practice the crystalline units are very small and are imperfectly aligned with many faults and defects. Some idea of the structure can be obtained from the schematic representation in Fig. 2.2 which is based on X-ray small angle scattering and electron microscopy studies. Full details can be found

Fig. 2.2. A schematic representation of the structure of carbon fibres based on X-ray diffraction and electron microscopy. (From S. C. Bennett, PhD thesis, University of Leeds 1976.)

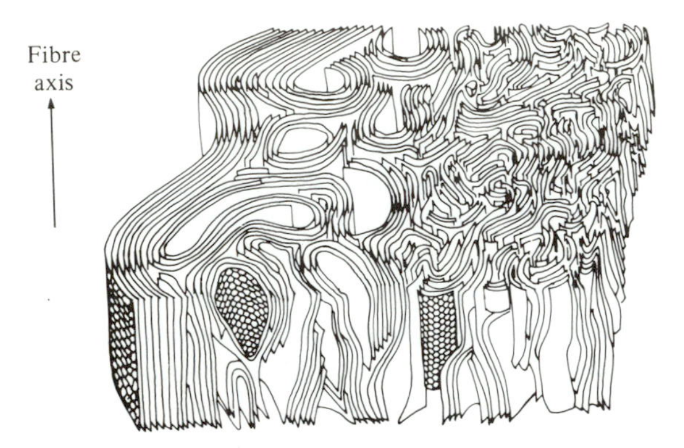

Fig. 2.3. Electron microscope lattice fringe image of a thin fragment of a Type I carbon fibre illustrating layer plane disorder. (From D. J. Johnson.)

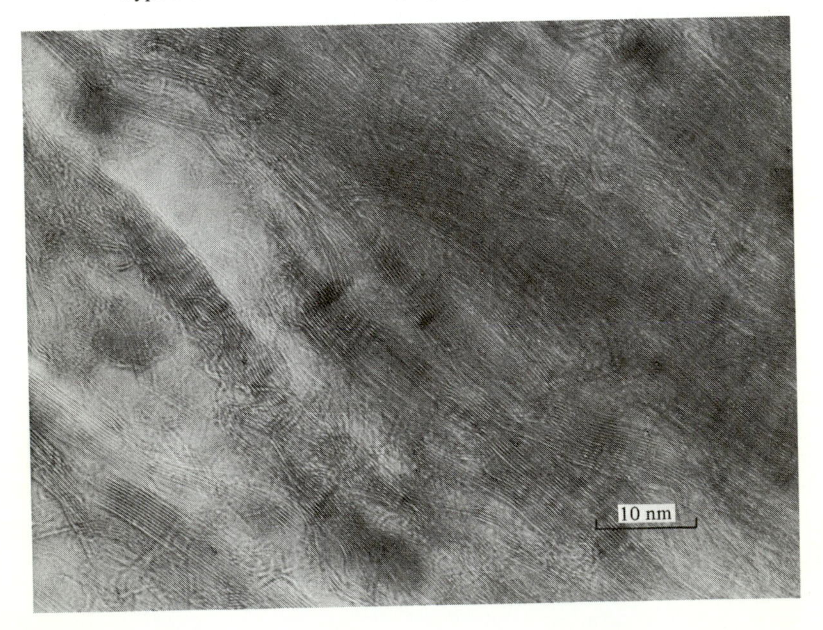

in reviews by Johnson (1971), Fourdeux, Perret & Ruland (1971), Reynolds (1973), and Johnson (1980). The average straight dimensions of the stacks of graphite layers parallel and perpendicular to the fibre axis and the thickness of the stack can be measured by X-ray diffraction and are typically less than 10 nm. An electron microscope lattice fringe image of the layer planes is shown in Fig. 2.3.

The modulus of carbon fibres depends on the degree of perfection of alignment which varies considerably with the manufacturing route and conditions. Imperfections in alignment, evident in Fig. 2.2, result in complex-shaped voids elongated parallel to the fibre axis. These act as stress raisers and points of weakness leading to a reduction in strength properties. Other sources of weakness, which are often associated with the manufacturing process, include surface pits and macrocrystallites. The effect of flaws has been discussed in detail by Reynolds & Moreton (1980).

The arrangement of the layer planes in the cross-section of the fibre is also important since it affects the transverse and shear properties of the fibre. Thus, for example, the normal polyacrylonitrile-based (PAN-based) Type I carbon fibres have a thin skin of circumferential layer planes and a core with random crystallites. In contrast some mesophase pitch-based fibres exhibit radially oriented layer structures. These different structures result in some significant differences in the properties of the fibres.

There are three main routes for producing fibres with the graphitic layers oriented preferentially parallel to the fibre axis. A reasonably full description is given by Gill (1972) and the methods can be summarised as follows:

(*a*) *Orientation of polymer precursor by stretching*. This process was developed by scientists at Rolls Royce and the Royal Aircraft Establishment, Farnborough, from original work by Shindo in Japan. It uses PAN as a starting material. PAN is a polymer closely resembling polyethylene in molecular conformation in which every alternate hydrogen side group of the polyethylene is replaced by a nitrile (–C≡N) grouping. In the first stage of the process the bulk PAN is converted into a fibre which is then stretched to produce alignment of the molecular chains along the fibre axis. When the stretched fibre is heated the active nitrile groups interact and produce a ladder polymer consisting of a row of six membered rings (Fig. 2.4). While the fibre is still under tension it is heated in an oxygen environment which leads to further chemical reaction and the formation of cross-links between the ladder molecules. The oxidised PAN is then reduced to give the carbon ring structure which is converted to 'turbostratic' graphite by heating at higher temperatures. The modulus and strength of the fibres depends on the final heat treatment temperature which determines the crystallite size and alignment. A similar process has been developed using rayon as the precursor.

(*b*) *Orientation by spinning*. This method, which was pioneered by Otani and his colleagues in Japan, involves melt spinning of molten pitch to produce fibres. Petroleum or coal tar pitch is thermally treated to temperatures above 350 °C to convert it by polymerisation to a 'mesophase pitch' which contains both isotropic and anisotropic (liquid crystal phase) material. The mesophase pitch is melt spun through a multihole spinneret to produce 'green' yarn. During this spinning process the hydrodynamic effects in the orifice result in orientation of the planar molecules. Different kinds of orientation can be induced depending on spinning conditions. The yarn is made infusible by oxidation at temperatures below its softening point to avoid fusing the filaments together. It is then carburised at temperatures normally around 2000 °C. Tensile stresses are required during some of these stages to prevent relaxation and loss of preferred orientation. The final crystallite size of mesophase pitch-based fibres is usually greater than PAN-based or rayon-based fibres.

(*c*) *Orientation during graphitisation*. At very high temperatures carburised fibres made from rayon, pitch or PAN as described above can be stretched during the graphitisation stage. This results in the sliding of the graphite layers over each other and further orientation of the layers parallel to the fibre axis.

All these processes are amenable to further development and increased sophistication. Each has its own advantages and disadvantages according

Fig. 2.4. Change of flexible polyacrylonitrile molecule into a rigid ladder molecule.

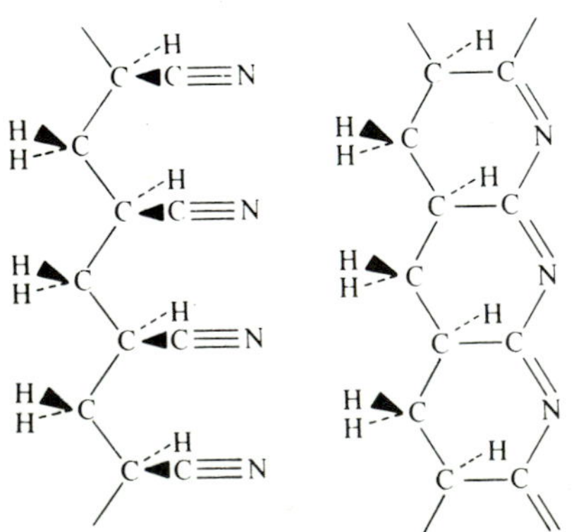

to performance, cost, ease of manufacture and quality of product required. By selection of the process and process variables the arrangement of the layers and perfection of alignment can be modified and the physical properties can be optimised to meet specific application requirements. Two examples will suffice. The properties of so-called Type I and Type II high modulus fibres made by the PAN process described in (*a*) above are listed in Table 2.1. Type I fibres have been graphitised to give maximum stiffness but have a relatively low strength, whereas Type II fibres have been graphitised to produce maximum strength. The variation of strength and stiffness with final graphitisation temperature is shown in Fig. 2.5, which is based on the pioneering work of Watt, Phillips & Johnson (1966).

The tensile modulus of Type I fibres of 390 GN m^{-2} is approximately 40% of the modulus of graphite crystals parallel to the *a*-directions, see Fig. 2.1. Thus, potentially higher moduli are possible if the perfection of the fibres is improved. The tensile strength of Type II fibres of 2.7 GN m^{-2} is well below the theoretical strength. This is due to poor alignment and flaws introduced during manufacture. Carbon fibres are almost completely brittle and show 100% elastic recovery when loaded to stresses below the fracture strength. The stress–strain curves of Type I and Type II fibres in Fig. 2.6 show that the high modulus fibres have a much lower strain to failure (0.50%) compared with the high strength

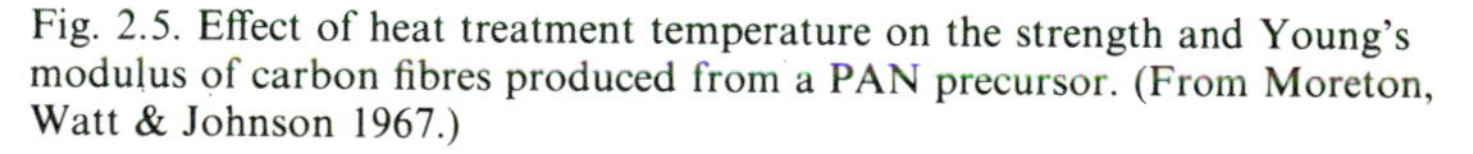
Fig. 2.5. Effect of heat treatment temperature on the strength and Young's modulus of carbon fibres produced from a PAN precursor. (From Moreton, Watt & Johnson 1967.)

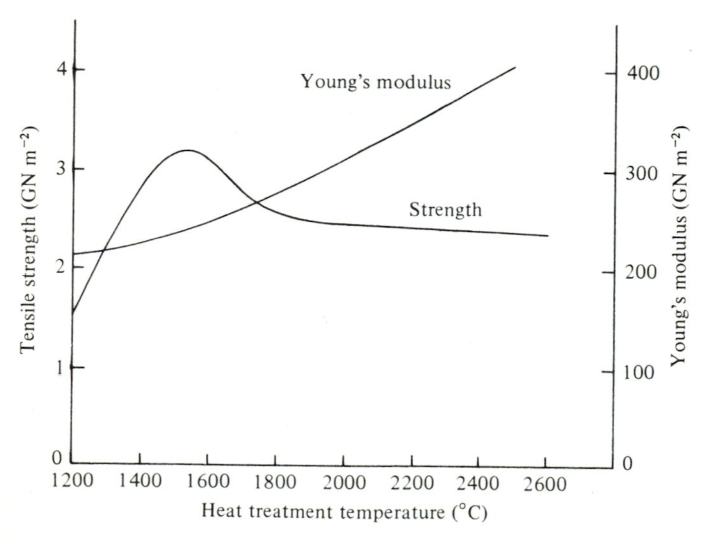

Table 2.1. *Properties of carbon, glass and Kevlar 49 fibres at* 20 °C

Property	Units	Carbon PAN-based Type I	Carbon PAN-based Type II	E glass	Aromatic Polyamide Kevlar 49
Diameter	μm	7.0–9.7	7.6–8.6	8–14	11.9
Density	10^3 kg m^{-3}	1.95	1.75	2.56	1.45
Young's modulus	GN m^{-2}	390	250	76	125
Modulus (perpendicular to fibre axis)	GN m^{-2}	12	20	76	
Tensile strength	GN m^{-2}	2.2	2.7	1.4–2.5 (typical) 3.5 (freshly drawn)	2.8–3.6
Elongation to fracture	%	0.5	1.0	1.8–3.2 (typical)	2.2–2.8
Coefficient of thermal expansion (0 to 100 °C)	10^{-6} °C^{-1}	−0.5 to −1.2 (parallel) 7–12 (radial)	−0.1 to −0.5 (parallel) 7–12 (radial)	4.9	−2 (parallel) 59 (radial)
Thermal conductivity (parallel to fibre axis)	W m^{-1} °C^{-1}	105	24	1.04	0.04

Notes: 1. Density of graphite single crystals is 2.26×10^3 kg m^{-3}.
2. Most of the information is obtained from manufacturer's data sheets; a wide range of values has been published and this information should only be used as a rough guide.

fibres (1.0%). As expected from a consideration of the microstructure of the fibres and, in particular, the orientation of the layer planes, the transverse tensile modulus and the longitudinal shear modulus are much smaller than the longitudinal tensile modulus and are dependent on the layer structure.

An important aspect of the mechanical properties of fibres is variability. Tensile tests on individual fibres show a wide range of strengths. For example, Proctor (1972) reported that in a single batch of Type I fibres the strength varied between 0.5 and 4.3 GN m^{-2} and Young's modulus varied between 270 and 580 GN m^{-2}. Similarly, tensile tests on fibres show that the measured average strength is strongly dependent on the test section gauge length varying, for example, from 2.1 GN m^{-2} for a 1 cm gauge length to 1.1 GN m^{-2} for a 20 cm gauge length. This type of variation is typical of fibres containing a distribution of flaws along their length since the probability of a flaw of a given length occurring in the test section increases with increasing gauge length. As with all brittle materials, it is necessary to

Fig. 2.6. Stress–strain curves of fibres. The vertical arrows indicate complete failure.

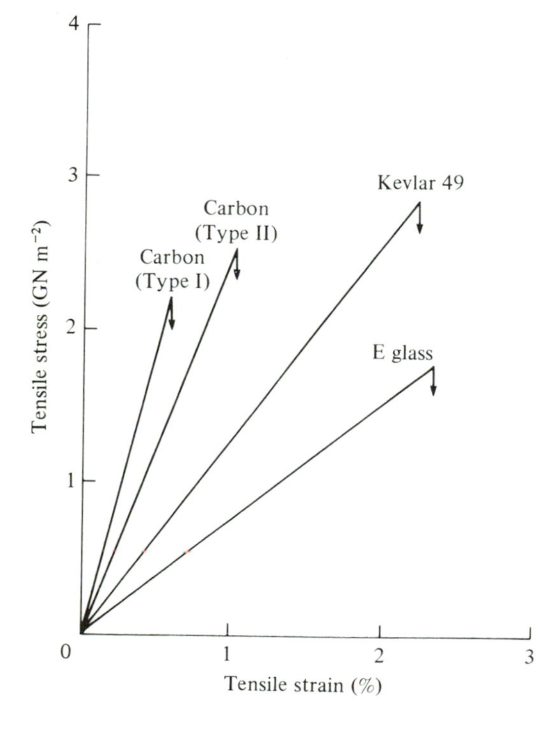

use a statistical approach to account for the strength of fibres and fibre bundles and this topic is considered in more detail in Section 7.2. For an example of recent work on the dependence of the strength of fibres on length using a statistical analysis see Hitchon & Phillips (1979).

2.2 Glass fibres

Many different compositions of mineral glasses have been used to produce fibres. The most common are based on silica (SiO_2) with additions of oxides of calcium, boron, sodium, iron and aluminium. These glasses are usually amorphous although some crystallisation may occur after prolonged heating at high temperatures. This usually leads to a reduction in strength properties. Typical compositions of the three well-known glasses used for glass fibre in composite materials are given in Table 2.2.

E glass (E for electrical) is the most commonly used glass because it draws well and has good strength, stiffness, electrical and weathering properties. C glass (C for corrosion) has a higher resistance to chemical corrosion than E glass but is more expensive and has lower strength properties. S glass is more expensive than E glass but has a higher Young's modulus and is more temperature resistant. It is used in special applications such as the aircraft industry where the higher modulus may justify the extra cost.

Continuous filament glass fibres are produced by melting the raw materials in a reservoir or tank which feeds the molten glass into a series of platinum bushings each of which has several hundred holes in its base. The glass flows under gravity and fine filaments are drawn mechanically downwards as the glass exudes from the holes. The glass fibres are wound onto drums at speeds of several thousand metres per minute.

Table 2.2. *Composition of glass used for fibre manufacture (all values in wt %)*

	E glass	C glass	S glass
SiO_2	52.4	64.4	64.4
Al_2O_3, Fe_2O_3	14.4	4.1	25.0
CaO	17.2	13.4	—
MgO	4.6	3.3	10.3
Na_2O, K_2O	0.8	9.6	0.3
Ba_2O_3	10.6	4.7	—
BaO	—	0.9	—

Note: Data from Fibreglass Ltd.

Control of the fibre diameter is achieved by adjusting the head of glass in the tank, the viscosity of the glass, which depends on composition and temperature, the diameter of the holes and the winding speed. The diameter of commercial E glass fibres is usually between 8 and 15 μm and is normally about 11 μm.

The strength and modulus of glass is determined primarily by the three-dimensional structure of the constituent oxides. Some idea of the amorphous structure can be obtained from Fig. 2.7 which shows a two-dimensional representation of the three-dimensional network of linked polyhedron units in a simple sodium silicate glass. Each polyhedron is a combination of oxygen atoms around a silicon atom and they are bonded to each other by strong covalent bonds. The sodium ions form ionic bonds with oxygen atoms and are not linked directly to the network. The structure of the network and the strength of the individual bonds can be varied by the addition of other metal oxides (see Table 2.2) and so it is possible to produce glass fibres with different chemical and physical properties. In contrast to carbon and Kevlar 49 fibres the properties of glass fibres are isotropic so that, for example, Young's modulus along the fibre axis is the same as transverse to the axis. This is a direct consequence of the three-dimensional network structure of the glass.

Fig. 2.7. Two-dimensional representation of the polyhedron network structure of sodium silicate glass.

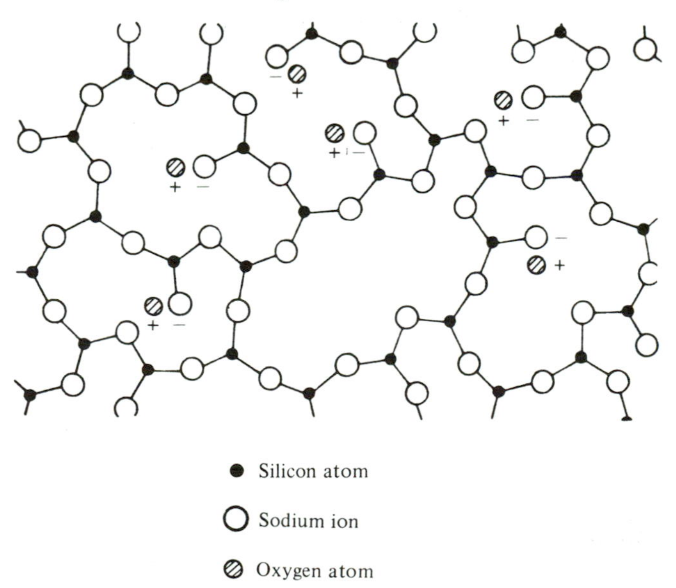

The properties of E glass fibre are listed in Table 2.1. The strength is strongly dependent on processing conditions and testing environment. Freshly drawn fibres which have been handled very carefully to avoid surface damage have a strength of 3.5 GN m^{-2} and the variability in strength is small. The strength decreases when the fibres are tested in humid air owing to absorption of water on the glass surface and there is a more marked decrease when the surface has been in contact with mineral acids. The strength and modulus is temperature dependent and glass is subject to static fatigue.

The most important factor determining the ultimate strength of glass is the damage which fibres sustain when they rub against each other during processing operations. The application of a size coating at a very early stage in manufacture helps to minimise the damage. The strengths quoted in Table 2.1 are for fibres produced commercially. Some damage is unavoidable and is produced randomly along the fibres. This leads to considerable variability in tensile strength and the same sort of gauge length dependence mentioned in Section 2.1 for carbon fibres. These points are well illustrated in Fig. 2.8 which compares the strength of single glass fibres extracted from a strand of fibres with the strength of freshly drawn virgin fibres. The mechanical damage is in the form of fine surface cracks. The effects of the cracks can be minimised or

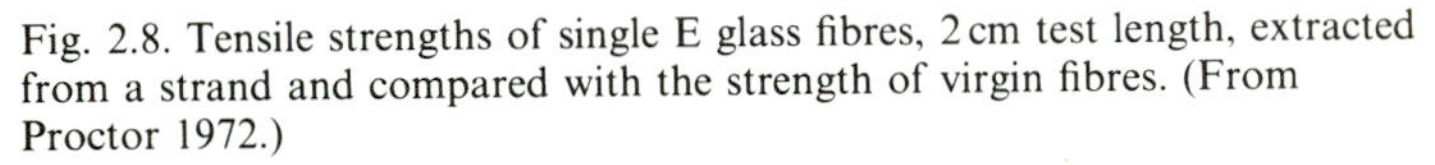

Fig. 2.8. Tensile strengths of single E glass fibres, 2 cm test length, extracted from a strand and compared with the strength of virgin fibres. (From Proctor 1972.)

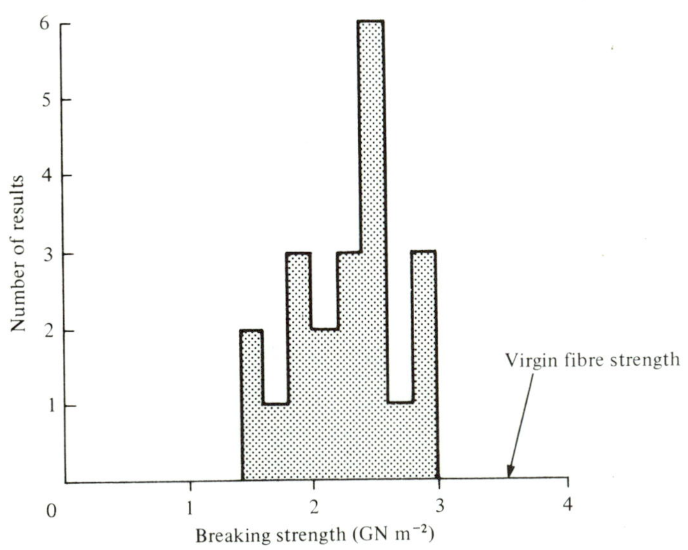

removed by surface etching but this is not a commercial proposition. A review of the sources of weakness in glass has been written by Gurney (1964).

The size is usually applied as a very thin coating to the fibres by spraying them with water containing an emulsified polymer. The size used depends on the future application of the fibres and serves several purposes: (*a*) to protect the surface of the fibres from damage, (*b*) to bind the fibres together for ease of processing, (*c*) to lubricate the fibres so that they can withstand abrasive tensions during subsequent processing operations through guide eyes and tensioning rollers, (*d*) to impart anti-static properties, (*e*) to provide a chemical link between the glass surface and the matrix to increase the interface bond strength as described in detail in Chapter 3. It is essential that the size is compatible with the matrix.

2.3 Organic fibres

A relatively new, and potentially important, class of fibres is based on the high strength and stiffness which is possible in fully aligned polymers. Frank (1970) has provided an excellent review of the physical principles involved. Thus, for example, chain extended polyethylene single crystals consist of straight zig-zag carbon to carbon bonded chains fully aligned and closely packed which have a theoretical modulus of about 220 GN m^{-2}. It is not yet possible to produce fibres with a fully aligned structure but extensive work by Ward and his colleagues (see Ward 1980) has led to the production of highly drawn high density polyethylene fibres with a tensile modulus of 60 GN m^{-2} and a tensile strength of 1.3 GN m^{-2}. Chain alignment and extension occurs during drawing and stretching. The difference between the theoretical maximum and the experimental values is due to imperfect alignment, chain folding and chain ends. As in carbon fibres the modulus normal to the fibres is less than that parallel to the fibres. The most successful commercial organic fibre to date has been developed by the Du Pont company with the trade name Kevlar. There is little detail available about the manufacturing processes but the initial patent in 1968 claims that poly-para-benzamide fibres can be produced with a modulus of 130 GN m^{-2} by a solvent spinning process.

Two forms, Kevlar 29 and Kevlar 49, are available, the latter was originally known by its development code number PDR 49. Kevlar 29 was developed for tyre cord reinforcement and has a high strength and intermediate modulus. Kevlar 49 has a higher modulus but the same strength as Kevlar 29 and is the preferred fibre for high performance composite materials.

Present day Kevlar is thought to be an aromatic polyamide called poly(paraphenylene terephthalamide) with the chemical formula below.

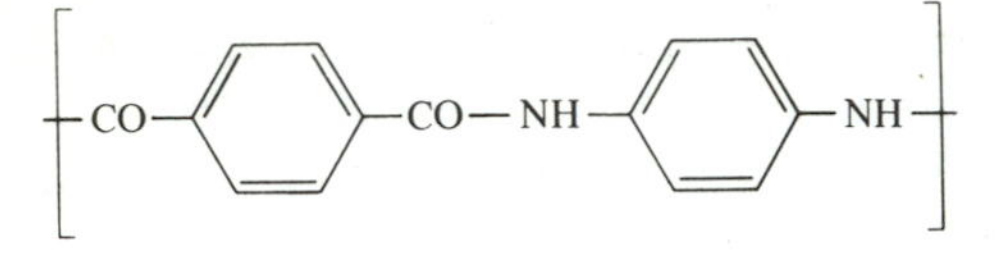

The aromatic rings result in the polymer molecules having the properties of a fairly rigid chain. The fibres are produced by extrusion and spinning processes. A solution of the polymer in a suitable solvent is held at a temperature between −50 °C and −80 °C before extrusion into a hot walled cylinder at 200 °C, whereupon the solvent evaporates. The resulting fibre is wound onto a suitable device. At this stage the fibre has a tensile strength of 840 MN m^{-2} and a tensile modulus of 3.6 GN m^{-2}. The fibre is then given a stretching and drawing treatment which increases the strength and stiffness properties to those of the finished fibre. Some typical properties are given in Table 2.1.

The processes used to manufacture the fibres result in alignment of the stiff polymer molecules parallel to the fibre axis and the modulus of 130 GN m^{-2} indicates that a high degree of alignment is achieved. The supramolecular structure has been investigated by Johnson and his

Fig. 2.9. Schematic representation of the supramolecular structure of aromatic polyamide fibres (Kevlar 49) depicting the radially arranged pleated system. (From Dobb, Johnson & Saville 1980.)

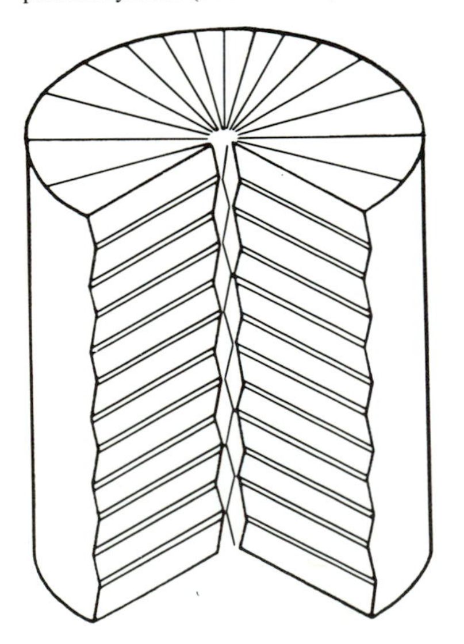

co-workers (see Dobb, Johnson & Saville 1980) by electron diffraction and dark field electron microscopy. A schematic representation of the structure is shown in Fig. 2.9. The molecules form rigid planar sheets (see Fig. 2.10) with the chain-extended molecules hydrogen bonded together. The sheets are stacked to form a crystalline array but there is only weak bonding between the sheets. As shown in Fig. 2.9 the sheets are arranged in the form of a radial system of axially pleated lamellae. Each component of the pleat is about 500 nm long and they are separated by short transitional bands. The angle between adjacent components of the pleat is about 170°. Fibres with this structure are likely to have a low longitudinal shear modulus and poor transverse properties. In particular, Kevlar fibres have a very low resistance to axial compressive failure. Aromatic polyamides have remarkably good resistance to temperature, given that they are organic polymers. Thus,

Fig. 2.10. Planar array of poly-*p*-phenylene terephthalamide molecules showing interchain hydrogen bonding. (From Dobb, Johnson & Saville 1980.)

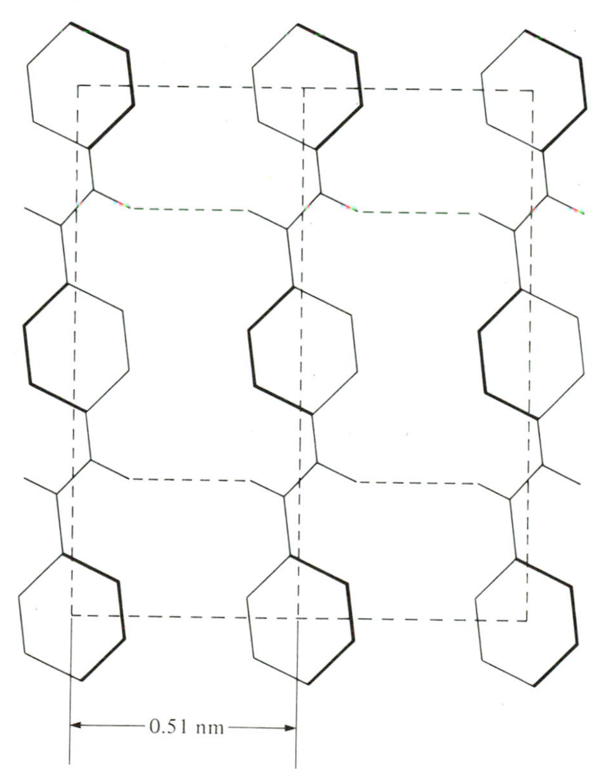

for example, there is only a 12% reduction in strength after 10 h at 250 °C and a 5% reduction after 500 h at 160 °C.

2.4 Comparison of fibres

The summary of properties in Table 2.1 reveals some important differences between carbon, glass and Kevlar 49 fibres which arise because of the different bonding configurations and types. In this section four aspects of fibre properties particularly relevant to the use of fibres in composite materials are considered.

Specific properties

Perhaps the most important factor leading to the rapid development of composite materials is the savings in mass which can be achieved from the use of low density fibres with high modulus and strength. The specific Young's modulus and specific tensile strength are defined as the measured modulus and strength divided by the density respectively. Some values are given in Table 2.3 based on the data in Table 2.1. Carbon fibres are far superior to E glass fibres in terms of specific modulus with Kevlar 49 fibres significantly better than E glass but inferior to carbon. Kevlar 49 has a specific tensile strength 40–50% greater than E glass and Type II carbon fibres. For comparison the specific modulus of a high strength plain carbon steel is 27 and the specific strength is 0.2 (same units as in Table 2.3). The comparison is not entirely justified because the fibres have to be incorporated into a resin to make a structural component. The maximum volume fraction of fibres normally achieved in aligned fibre composites is about 0.70 so that the values have to be reduced by this factor. It must be remembered also that the mechanical properties of aligned composite materials transverse to the fibres are much less than those parallel to the fibres.

Thermal stability

In the absence of air and other oxidising atmospheres carbon fibres possess exceptionally good high temperature properties. Indeed, reference to Fig. 2.5 indicates that the strength and modulus depend on the final heat treatment temperature which is in the range 1200 to 2600 °C. Carbon fibres retain their superior properties well above 2000 °C. For applications involving polymer matrices this property cannot be used because most matrices lose their properties above about 200 °C.

Bulk glass has a softening temperature of about 850 °C but the strength and modulus of E glass fibres decreases rapidly above 250 °C. Although the thermal stability of Kevlar 49 is inferior to both glass and

Table 2.3. *Specific Young's modulus and specific tensile strength*

	Units	Carbon PAN-based Type I	Carbon PAN-based Type II	E glass	Kevlar 49
Specific Young's modulus (modulus/density)	$GN\,m^{-2}/10^3\,kg\,m^{-3}$	200	143	30	86
Specific tensile strength (strength/density)	$GN\,m^{-2}/10^3\,kg\,m^{-3}$	1.1	1.5	1.4 (freshly drawn)	2.2

carbon, it is probably adequate for use in most polymer matrix systems. Apart from retaining the properties during use in service at elevated temperatures, it is essential that deterioration in properties does not occur during manufacturing operations. The properties of glass are likely to be nearly reversible with temperature but Kevlar 49 will suffer irreversible deterioration owing to changes in internal structure. Heating during processing occurs in curing of thermosetting resins and in the melt processing of thermoplastics. Thus, the injection moulding temperatures of nylon and polycarbonate is usually in the range 270–320 °C.

Kevlar fibres undergo severe photodegradation on exposure to sunlight. Both visible as well as ultraviolet light have an effect and lead to discoloration and reduction in mechanical properties. The degradation can be avoided by coating the surface of the composite material with a light absorbing layer.

Compressive properties

No reference is made in Table 2.1 to the axial compressive stiffness and strength of fibres. These properties are difficult to measure and can only be inferred from the properties of composite materials fabricated with the fibres. It is found that the axial stiffness in compression is approximately the same as the stiffness in tension for all the fibres. However, the data in Table 7.1 indicate that the longitudinal compressive strength of aligned unidirectional laminae made from Kevlar is only 20% of the tensile strength. In contrast the corresponding data for glass and carbon fibres show that they have similar strengths in tension and compression.

The low compressive strength of Kevlar is due to the anisotropic properties of the fibre and the low shear stiffness. Basically, like other textile fibres, the material has only limited elasticity in compression. In tension the loads are carried by the covalent bonds but in compression the weak hydrogen bonding and van der Waal forces mean that local yielding and fibrillation can occur. This leads to buckling and kink formation in the fibres and is referred to below in connection with fibre flexibility.

Fibre fracture and flexibility

The stress–strain curves shown in Fig. 2.6 suggest that in tension all the fibres fracture in a brittle manner without any yield or flow. Carbon and glass fibres are almost completely brittle and fracture without any reduction in the cross-sectional area (Fig. 2.11). In contrast Kevlar 49 fibres fracture in a ductile manner. Pronounced necking precedes fracture and final separation occurs after a large amount of local drawing. Fracture often involves fibrillation of the fibres and some features of the fibrillar microstructure are evident on the fibre surface in Fig. 2.11*b*.

The diameter of fibres has a large effect on the ease with which they can be bent. This is important in operations where fibres are fed through eyes and over bobbins as in weaving and filament winding and in moulding and mixing operations where fibres are intimately mixed with polymer in high shear mixers and subsequently extruded or injection

Fig. 2.11. Scanning electron micrographs of fibres fractured in tension, (*a*) carbon, (*b*) Kevlar 49, (*c*) glass.

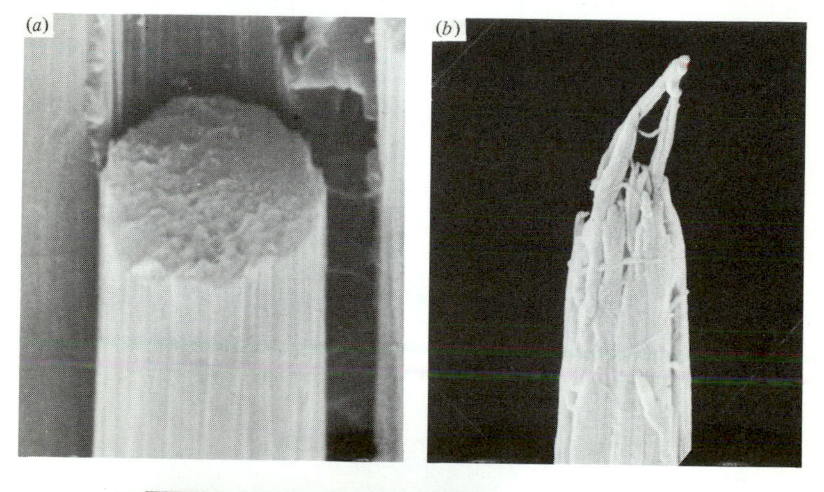

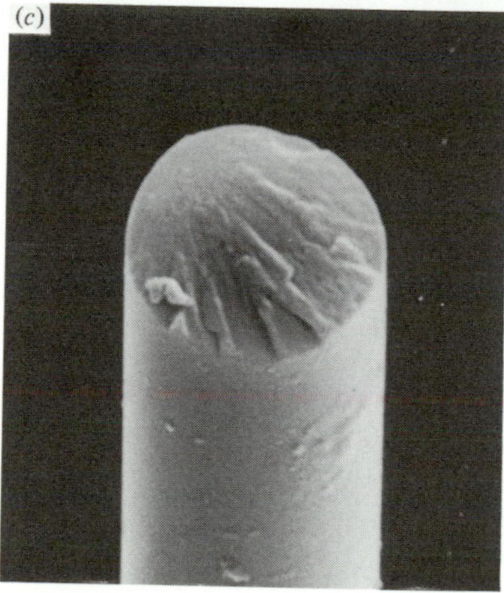

moulded using screw feed machines. Insufficient flexibility leads to difficulties in handling fibres and in fibre breakage.

The flexibility of a fibre can be expressed in terms of the moment M required to bend a fibre with a circular cross-section to a given radius of curvature ρ

$$M = \pi E d^4/64\rho \qquad (2.1)$$

where E is Young's modulus and d is the fibre diameter. The flexibility $1/M\rho$ is dominated by d but also depends on E. A comparison of the flexibilities of carbon, glass and Kevlar 49 is given in Table 2.4. The flexibilities of Type II carbon fibres and E glass fibres are significantly larger than Type I carbon fibres because of the lower modulus of these fibres.

Bending of fibres results in high surface tensile stresses which lead to fibre fracture. Assuming that the elastic deformation of the fibres is linear, the surface tensile stress is given by

$$\sigma = Ed/2\rho \qquad (2.2)$$

where ρ is the radius of curvature of the fibre when it is bent. For a given fibre fracture strength σ_f there will be a minimum radius of curvature which the fibres can sustain before fracture occurs, namely

$$\rho_{min} = Ed/2\sigma_f \qquad (2.3)$$

Values of ρ_{min} are given in Table 2.4 neglecting the statistical variation in fracture strength along the fibre. On this basis freshly drawn E glass fibres are much less susceptible to fibre breakage by bending than Type I carbon fibres with Type II carbon fibres in an intermediate position.

Table 2.4. *Flexibility of carbon, glass and Kevlar 49 fibres*

	Units	Type I carbon	Type II carbon	E glass	Kevlar 49
Diameter, d	μm	8	8	11	12
Modulus, E	$GN\,m^{-2}$	390	250	76	130
Flexibility ratio, Type I carbon fibre = 1	—	1.00	1.56	1.44	0.59
Fracture strength, σ_f	$GN\,m^{-2}$	2.2	2.7	3.5	
Minimum radius of curvature, ρ_{min}	mm	0.71	0.37	0.12	

Kevlar 49 fibres do not fit into this simple pattern because of the low compressive strength mentioned above. Bending of the fibres produces high surface compressive stresses as well as tensile stresses. Long before the bending curvature is sufficient to cause tensile fracture, the compressive region of the fibre undergoes yielding by the development of deformation bands. This results in a permanent deformation, as illustrated in Fig. 2.12.

An example of the break-down of glass fibres during moulding, as revealed by measurements of fibre length distribution is given in Section 4.6. In addition to bending stresses, high axial tensile stresses may develop in the moulding operations which increase the probability of fibre fracture. Glass fibres are also subject to abrasive damage which occurs when fibres rub together during processing and this leads to a reduction in σ_f and hence an increase in ρ_{min}. The bending of fibres during deformation and fracture of composite materials, particularly at the tip of propagating cracks is another area where fibre diameter and strength can affect properties.

Fig. 2.12. Scanning electron micrograph of a Kevlar 49 fibre showing deformation bands on the compression side of a sharp bend.

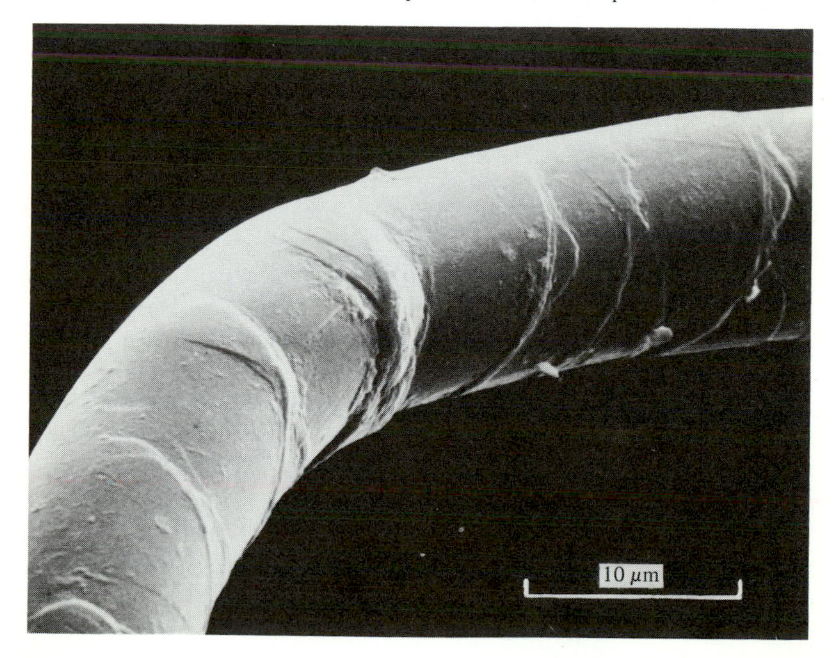

2.5 Thermosetting resins

Epoxy and polyester resins cover a very broad class of chemicals and a corresponding range of physical and mechanical properties can be obtained. This section is concerned primarily with final properties but a brief mention of the curing process is relevant. In thermosetting polymers, the liquid resins are converted into hard brittle solids by chemical cross-linking which leads to the formation of a tightly bound three-dimensional network of polymer chains. The mechanical properties depend on the molecular units making up the network and on the length and density of the cross-links. The former is determined by the initial chemicals used and the latter by the control of the cross-linking processes which are involved in the cure. Curing can be achieved at room temperature but it is usual to use a cure schedule which involves heating at one or more temperatures for pre-determined times to achieve optimum cross-linking and hence optimum properties. A relatively high temperature final post-cure treatment is often given to minimise any further cure and change in properties during service. Shrinkage during curing and thermal contraction on cooling after cure can lead to built-in stresses in composite materials (see, for example, Section 5.6).

The properties of the cured resins can be determined from specimens prepared by casting the uncured resins into moulds. A set of typical properties for epoxy and polyester resins is given in Table 2.5 which was compiled by Johnson (1979) from manufacturer's literature. These properties are subject to wide variations depending on the chemical system used and the cure conditions.

Thermosetting resins are usually isotropic. Their most characteristic property is in response to heat since, unlike thermoplastics, they do not melt on heating. However, they lose their stiffness properties at the heat distortion temperature and this defines an effective upper limit for their use in structural components. Epoxy resins are generally superior to polyester resins in this respect but other resins are available which are stable at higher temperatures such as aromatic polyamides and polyimides. Table 2.5 shows that epoxy resins have superior strength and elastic properties with a lower shrinkage on curing and a lower coefficient of thermal expansion. The strength of the interface bond between resin and fibre is also higher for epoxy resins. However, they have the disadvantage of a higher viscosity before curing and they are more expensive.

Thermosetting resins are normally regarded as brittle solids. This statement must be qualified since the brittleness evident in simple uniaxial tensile tests is due, in part, to specimen preparation and test procedures. This is best illustrated by reference to Fig. 2.13. The full

Table 2.5. *Comparison of typical properties of epoxy and polyester resins used in composite materials (after Johnson 1979)*

Property	Units	Epoxy resins	Polyester resins
Density	$Mg\,m^{-3}$	1.1–1.4	1.2–1.5
Young's modulus	$GN\,m^{-2}$	3–6	2–4.5
Poisson's ratio		0.38–0.4	0.37–0.39
Tensile strength	$MN\,m^{-2}$	35–100	40–90
Compressive strength	$MN\,m^{-2}$	100–200	90–250
Elongation to break (tension)	%	1–6	2
Thermal conductivity	$W\,m^{-1}\,°C$	0.1	0.2
Coefficient of thermal expansion	$10^{-6}\,°C^{-1}$	60	100–200
Heat distortion temperature	°C	50–300	50–110
Shrinkage on curing	%	1–2	4–8
Water absorption 24 h to 20 °C	%	0.1–0.4	0.1–0.3

Fig. 2.13. Stress–strain curves of a general purpose grade polyester resin, tested in tension and compression. The broken line is predicted by assuming $\sigma_{YC}/\sigma_{YT} = 1.3$, where σ_{YC} and σ_{YT} are the yield strengths in compression and tension respectively.

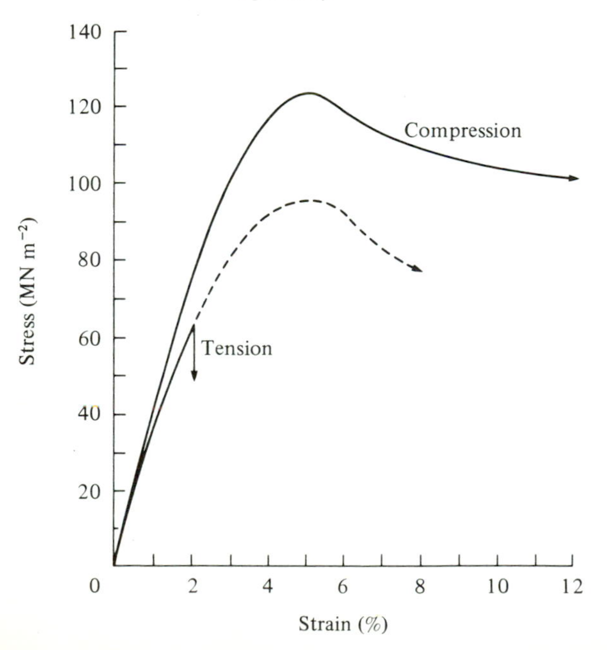

lines are the experimental stress–strain curves of a polyester resin tested in uniaxial tension and compression. In tension fracture occurs at 63 $MN\ m^{-2}$ at a strain of 2.0% and there is no sign of yielding before fracture. In compression, large scale yielding occurs with an upper yield strength of 122 $MN\ m^{-2}$. The plastic deformation processes evident in compression are suppressed in tension by premature fracture which is associated with flaws in the material and on the surface of the test specimens. In plastics which yield in tension the ratio of the yield strength in compression to the yield strength in tension $\sigma_{YC}/\sigma_{YT} \approx 1.3$. The broken line in Fig. 2.13 is the predicted stress–strain curve in tension when premature fracture is avoided.

The yield strength of plastics is strongly dependent on biaxial stress conditions, and Fig. 2.14 shows that the locus of the yield strength in σ_1, σ_2-space approximates to

$$\sigma_1^2+\sigma_2^2+(\sigma_1+\sigma_2)(\sigma_{YC}-\sigma_{YT})-\sigma_1\sigma_2 = \sigma_{YC}\sigma_{YT} \tag{2.4}$$

This is a modified form of the von Mises' yield criterion which takes account of the difference in σ_{YC} and σ_{YT} and the influence of the mean normal stress. σ_1 and σ_2 are principal stresses and σ_{YC} and σ_{YT} are the absolute values of the compressive and tensile yield stresses respectively. Equation (2.4) can be normalised by introducing ratios

$$R_1 = \sigma_1/\sigma_{YT} \quad \text{and} \quad R_2 = \sigma_2/\sigma_{YT}$$

and the equation becomes

$$R_1^2+R_2^2-R_1R_2+(R_1+R_2)\left(\frac{\sigma_{YC}}{\sigma_{YT}}-1\right) = \frac{\sigma_{YC}}{\sigma_{YT}} \tag{2.5}$$

Fig. 2.14. Locus of yield strength of polymers in biaxial stress fields, calculated from equation (2.5) using $\sigma_{YC}/\sigma_{YT} = 1.3$. (From Caddell & Raghava 1976; see also Ward 1971.)

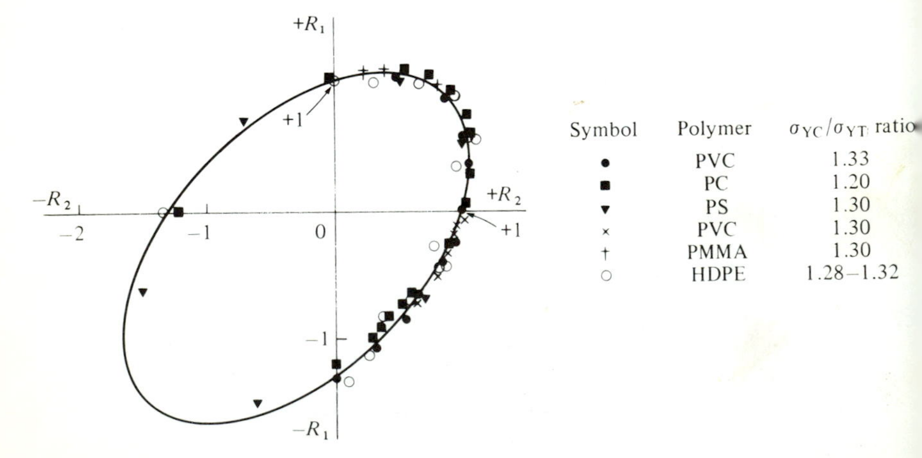

The results normalised in this way for a number of crystalline and amorphous polymers are shown in Fig. 2.14. A similar type of relation will apply for triaxial stress conditions but there are few experimental data available.

Resin and processes

There are three main routes to the manufacture of fibre reinforced composite materials with thermosetting resins. In the wet processing route low viscosity resin is impregnated into dry fibre. This is achieved in a number of ways: firstly, by wet lay-up in which the fibres in the form of a mat are impregnated with resin by rolling and pressing: secondly, by wet winding or filament winding in which the fibre tows, i.e. bundles of fibres, are drawn through a bath of resin before winding onto a mandrel or former of the required shape; thirdly, by resin injection, in which the fibres are placed in position in a closed mould and the resin is fed in under gravity or external pressure. The success of these processes depends on complete wetting of the fibres by the resin which, in turn, depends on the viscosity of the resin and the surface tension of the interface between the resin and the fibre. In composite materials with a large volume fraction of small diameter fibres the spaces between the fibres is very small (see Section 4.2) and wetting out requires long times and high pressures.

The second main route also involves wetting out of the fibres but fabrication occurs in two distinct stages. The first stage is the production of a pre-preg which is a tape or sheet of fibres impregnated with resin. This is produced by laying the fibres and resin between sheets of siliconised paper or plastic film which are pressed and rolled to ensure consolidation and wetting out of the fibres, and then partially cured to produce a flexible aggregate. The process allows excellent alignment of the fibres in unidirectional layers. The second stage, after removal of the paper or film, is to form the shape of the component by stacking up layers of pre-preg, or winding pre-preg tape onto a mandrel. The pre-preg is consolidated by pressure and the final cure is achieved by heating under pressure.

The third route is a variation of the pre-preg approach. The intermediate products are sheet moulding compound (SMC) and dough moulding compound (DMC) based on polyester resins. To make SMC, resin containing chemical thickening agents and particulate fillers such as calcium carbonate, is mixed with chopped fibres to form a sheet. The fibres lie mainly parallel to the sheet and the volume fraction is usually in the range 15 to 40%. DMC is made in a similar way but the material has a higher concentration of filler and a lower fibre content and the fibres are oriented in three dimensions. After mixing of fibres, resin and filler the material has the consistency of a dough. This is converted into a strong rigid material by hot pressing in moulds which results in further consolidation of the material and curing of the resin. Considerable amounts of flow are involved, particularly in DMC compression and SMC injection moulding processes, and this leads to fibre alignment (see Section 4.9). The amount of alignment may vary considerably in different parts of the mould and is an important factor in the design of components and the prediction of properties.

2.6 Thermoplastics

Unlike thermosetting resins, thermoplastics are not cross-linked. They derive their strength and stiffness from the inherent properties of the monomer units and the very high molecular weight. This ensures that in amorphous thermoplastics there is a high concentration of molecular entanglements, which act like cross-links, and in crystalline materials there is a high degree of molecular order and alignment. In amorphous materials heating leads to disentanglement and a change from a rigid solid to a viscous liquid. In crystalline materials heating results in melting of the crystalline phase to give an amorphous viscous liquid.

Both amorphous and crystalline polymers may have anisotropic properties depending on the conditions during solidification. In amorphous polymers this is due to molecular alignment which occurs during melt flow in moulding the material or subsequently during plastic deformation. Similarly, in crystalline polymers, the crystalline lamellar units can develop a preferred orientation due, for example, to non-uniform nucleation at surfaces or in the flowing melt and preferential growth in some directions because of temperature gradients in the melt. There is increasing evidence that these effects can have a profound effect on the properties of composite materials.

Thermoplastic matrices are normally used with short fibre reinforcement for applications in products made by injection moulding. The feed-stock is usually in the form of pellets which contain the short fibres, typically 1–3 mm long, intimately mixed and dispersed in the matrix. Three common matrix polymers are polypropylene, nylon and polycarbonate. The first two are usually semicrystalline with approximately 25–50% crystallinity and the latter is amorphous. Some typical properties are given in Table 2.6. In practice a wide range of properties can be obtained and the strength and heat resistance properties are particularly sensitive to processing history, molecular weight, molecular weight distribution and molecular chemistry. All these plastics yield and

Table 2.6. *Comparison of typical properties of three common thermoplastics used in composite materials at* 20 °C, *collected from manufacturer's literature*

Property	Units	Poly-propylene	Nylon 6.6	Poly-carbonate
Density	$Mg\,m^{-3}$	0.90	1.14	1.06–1.20
Young's modulus	$GN\,m^{-2}$	1.0–1.4	1.4–2.8	2.2–2.4
Poisson's ratio		0.3	0.3	0.3
Tensile yield strength	$MN\,m^{-2}$	25–38	60–75	45–70
Elongation to break (tension)	%	> 300	40–80	50–100
Thermal conductivity	$W\,m^{-1}\,°C$	0.2	0.2	0.2
Coefficient of thermal expansion	$10^{-6}\,°C^{-1}$	110	90	70
Melting point	°C	175	264	—
Deflection temperature under load at 1.82 $MN\,m^{-2}$	°C	60–65	75	110–140
Water absorption 24 h to 20 °C	%	0.03	1.3	0.1

undergo large deformations before final fracture and their mechanical properties are strongly dependent on the temperature and applied strain rate. Some indication of their relative sensitivity to temperature is given by the temperature of deflection under load which is measured in a three-point bending test. Another important feature of the properties, which is common to all thermoplastics, is that under constant load conditions the strain increases with time, i.e. the materials creep under load. This means that in composite systems there will be a redistribution of the load between resin and fibres during deformation.

References and further reading

Carbon fibres

Bacon, R. (1980) Carbon fibres from mesophase pitch. *Phil. Trans. R. Soc. Lond.* A **294**, 437–42.

Diefendorf, R. J. & Tokarsky, E. (1975) High performance carbon fibres. *Polym. Engng Sci.* **15**, 150–9.

Fourdeux, A., Perret, R. & Ruland, W. (1971) General structural features of carbon fibres. *Proceedings of the First International Conference on*

Carbon Fibres, their Composites and Applications, pp. 57–67. Plastics Institute, London.

Gill, R. M. (1972) *Carbon Fibres in Composite Materials*. Plastics Institute, London.

Hitchon, J. W. & Phillips, D. C. (1979) The dependence of the strength of carbon fibres on length. *Fibre Sci. Technol.* **12**, 217–33.

Johnson, D. J. (1971) Microstructure of various carbon fibres. *Proceedings of the First International Conference on Carbon Fibres, their Composites and Applications*, pp. 52–6. Plastics Institute, London.

Johnson, D. J. (1980) Recent advances in studies of carbon fibre structure. *Phil. Trans. R. Soc. Lond.* A **294**, 443–9.

Moreton, R., Watt, W. & Johnson, W. (1967) Carbon fibres of high strength and high breaking strain. *Nature, Lond.* **213**, 690–1.

Reynolds, W. N. (1973) Structure and physical properties of carbon fibres. *Chemistry and Physics of Carbon*, vol. 2, ed. P. L. Walker & P. A. Thrower, pp. 2–68. Marcel Dekker, Maidenhead, Berkshire.

Reynolds, W. N. & Moreton, R. (1980) Some factors affecting the strengths of carbon fibres. *Phil. Trans. R. Soc. Lond.* A **294**, 451–61.

Watt, W., Phillips, L. N. & Johnson, W. (1966) High-strength high-modulus carbon fibres. *The Engineer* **221**, 815–16.

Glass

Gurney, C. (1964) Sources of weakness in glass. *Proc. R. Soc. Lond.* A **282**, 24–33.

Lowrie, R. E. (1967) Glass fibres for high strength composites. *Modern Composite Materials*, ed. L. J. Broutman & R. H. Krock, ch. 11. Addison Wesley, Reading, Mass.

Proctor, B. A. (1972) Fibre reinforced composite materials. *Faraday Special Discussions of The Chemical Society*, no. 2, 63–76.

Proctor, B. A. (1980) Glass fibres for cement reinforcement. *Phil. Trans. R. Soc. Lond.* A **294**, 427–36.

Organic fibres

Dobb, M. G., Johnson, D. J. and Saville, B. P. (1980) Structural aspects of high modulus aromatic polyamide fibres. *Phil. Trans. R. Soc. Lond.* A **294**, 483–5.

Frank, F. C. (1970) The strength and stiffness of polymers. *Proc. R. Soc. Lond.* A **319**, 127–36.

Magat, E. E. (1980) Fibres from extended chain polyamides. *Phil. Trans. R. Soc. Lond.* A **294**, 463–72.

Ward, I. M. (1980) Ultra-high modulus polyolefins. *Phil. Trans. R. Soc. Lond.* A **294**, 473–82.

Polymer matrices and processing

Caddell, R. M. & Raghava, R. S. (1976) On the macroscopic yield behaviour of isotropic and anisotropic polymers. *Rev. Deform. Behav. Mater.* **1**, 207–27.

Johnson, A. F. (1979) *Engineering Design Properties of GRP*. British Plastics Federation, London.

Lee, H. & Neville, K. (1967) *Handbook of Epoxy Resins*. McGraw-Hill, New York.

Parkyn, B. (1967) *Polyesters*. Iliffe.

SPI Handbook of Technology and Engineering of Reinforced Plastics/composites (1973), ed. J. G. Mohr. van Nostrand Reinhold, New York.

Ward, I. M. (1971) Review: The yield behaviour of polymers. *J. Mater. Sci.* **6**, 1397–417.

Williams, J. G. (1973) *Stress Analysis of Polymers*. Longman, London.

3 Fibre–matrix interface

3.1 Introduction

The structure and properties of the fibre–matrix interface play a major role in the mechanical and physical properties of composite materials. In particular, the large differences between the elastic properties of the matrix and the fibres have to be communicated through the interface or, in other words, the stresses acting on the matrix are transmitted to the fibre across the interface. Consider the simple example illustrated in Fig. 3.1*a* in which the composite material is represented by alternate sheets of material with different elastic properties. In the absence of a chemical, physical or mechanical bond between the layers the composite has no tensile strength in direction AA' normal to the layer planes. The strength and modulus in direction BB', parallel to the layers, depends on the way the sample is gripped. If no bond exists and a simple adhesive grip is made to the outer layers (Fig. 3.1*b*) the strength is limited to the strength of the outer layers since the applied load is taken up entirely by these layers. On the other hand, if the layers are all clamped together in the grips (Fig. 3.1*c*) all the layers take the load and the composite will be much stronger and stiffer. It follows from this example that in order to use the high strength and stiffness of the fibres, they have to be strongly bonded to the matrix.

A theoretical analysis of the way stress is transferred from the fibre to the matrix is given in Section 5.4. In the analysis a number of assumptions have to be made about the properties of the interface. These are (i) the matrix and the fibre behave as elastic materials, (ii) the interface is infinitesimally thin, (iii) the bond between the fibre and matrix is perfect which implies that there is no strain discontinuity across the interface, (iv) the material close to the fibre has the same properties as the material in bulk form, and (v) the fibres are arranged in a regular or repeating array. These assumptions are necessary to obtain solutions to the mathematical models and the results obtained are an important guide to what is likely to happen in real composite systems. However, none of the assumptions (ii) to (iv) is strictly correct because the actual interface has a complicated chemical and physical structure. Assumptions (i) and (v) also require some qualification as will be evident from Chapters 2 and 4 respectively.

Some appreciation of the real properties of the interface is essential for an understanding of composite material properties. Thus, for example, the interface is a dominant factor in the fracture toughness properties of composite materials and in their response to aqueous and corrosive environments. Composite materials with weak interfaces have relatively low strength and stiffness but high resistance to fracture whereas materials with strong interfaces have high strength and stiffness but are very brittle. The effect is related to the ease of debonding and pull-out of fibres from the matrix during crack propagation.

In this chapter an insight is given into the nature of the bonding between the matrix and the fibre. Since this is dependent on the atomic arrangement and chemical properties of the fibre and on the molecular conformation and chemical constitution of the polymer matrix, it follows that the interface is specific to each fibre–matrix system. The

Fig. 3.1. Schematic diagram of a layer composite material to illustrate the importance of interface bond strength.

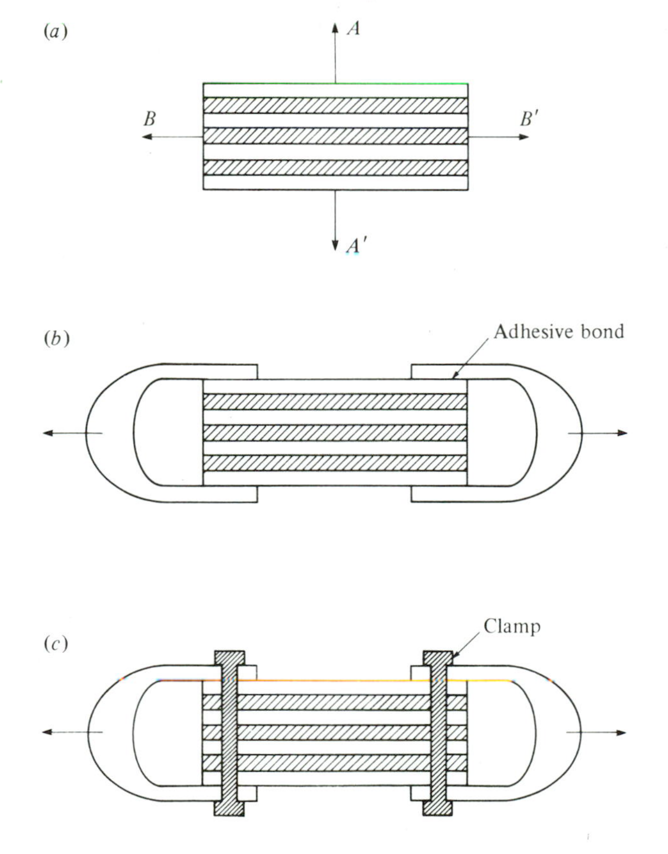

theories of adhesion are reviewed briefly to give some indication of the origin of bond strength. The specific case of the bonding of glass fibres to thermosetting resins is considered in more detail as an example to highlight the type of detail which is required. This is followed by a consideration of the concept of bond strength and an account of the methods which have been used to measure it and relate it to composite properties.

3.2 Theories of adhesion

In a simple system bonding at an interface is due to adhesion between fibre and matrix. However, the fibres are often coated with a layer of material which forms a bond between the fibre and matrix. For the purpose of this section this is an unnecessary complication. Adhesion can be attributed to five main mechanisms which can occur at the interface either in isolation or in combination to produce the bond (see Wake (1978) for a more detailed review).

(*a*) *Adsorption and wetting*. When two electrically neutral surfaces are brought sufficiently close together there is a physical attraction which is best understood by considering the wetting of solid surfaces by liquids. In the case of two solids being brought together the surface roughness on a micro and atomic scale prevents the surfaces coming into contact except at isolated points as illustrated in Fig. 3.2*a*. In addition the surfaces are usually contaminated. Even if the contamination is removed, and strong adhesion occurs at the contact points, the

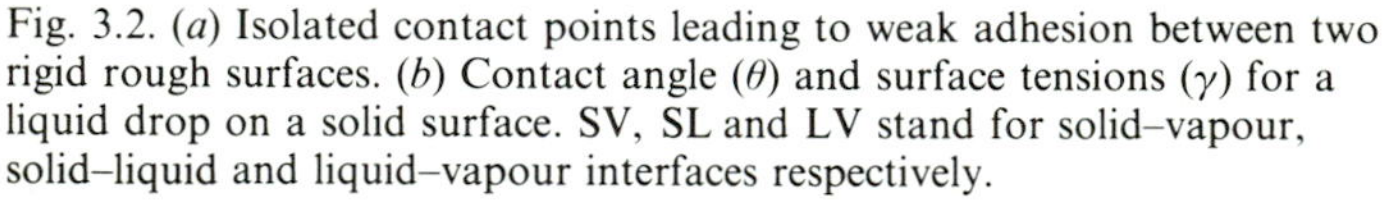

Fig. 3.2. (*a*) Isolated contact points leading to weak adhesion between two rigid rough surfaces. (*b*) Contact angle (θ) and surface tensions (γ) for a liquid drop on a solid surface. SV, SL and LV stand for solid–vapour, solid–liquid and liquid–vapour interfaces respectively.

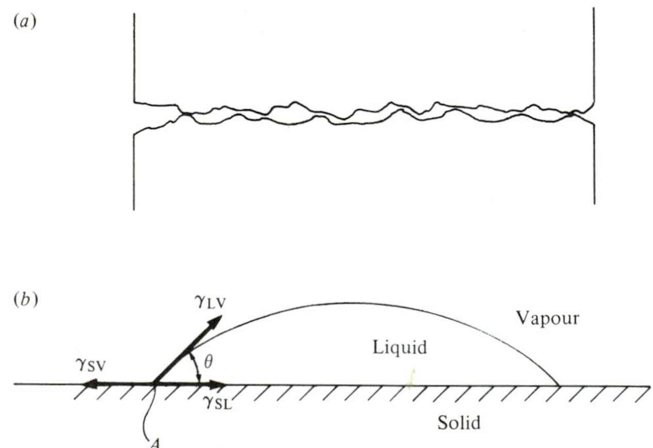

adhesion averaged over the whole surface will be weak. For effective wetting of a fibre surface the liquid resin must cover every hill and valley of the surface to displace all the air. Weak boundary layers must be avoided also.

Wetting can be understood in terms of two simple equations. The Dupré equation for the thermodynamic work of adhesion, W_A, of a liquid to a solid states that

$$W_A = \gamma_1 + \gamma_2 - \gamma_{12} \tag{3.1}$$

where γ_1 and γ_2 are the surface free energies of the liquid and solid respectively and γ_{12} is the free energy of the liquid–solid interface. This equation can be related to the physical situation of a liquid drop on a solid surface, as in Fig. 3.2*b*, by using the Young equation. When the forces at a point A are resolved in the horizontal direction, then Young's equation states

$$\gamma_{SV} = \gamma_{SL} + \gamma_{LV} \cos \theta \tag{3.2}$$

where γ_{SV}, γ_{SL} and γ_{LV} are the surface free energies, or surface tensions of the solid–vapour, solid–liquid and liquid–vapour interfaces respectively, and θ is the contact angle. For spontaneous wetting to occur $\theta = 0°$. The surface tension of solids is very difficult to measure whereas the surface tension of liquids can be determined relatively easily. A measure of γ_{SV} can be obtained from the way liquids of known γ_{LV} wet the solid. Zisman introduced the concept of critical surface tension of wetting γ_c such that only liquids with $\gamma_{LV} < \gamma_c$ will spontaneously spread on the solid. This is a useful parameter in considering the wetting of fibres by resins.

It follows that glass and graphite with theoretically calculated (Kelly 1973) surface energies of 560 mJ m^{-2} and 70 mJ m^{-2} respectively will be readily wetted by polyester and epoxy resins with surface energies of 35 mJ m^{-2} and 43 mJ m^{-2} respectively provided the viscosity of the resins is not too high. In contrast it will be difficult to wet polyethylene, which has a measured γ_c of 31 mJ m^{-2}, with these resins and some surface treatment will be required.

A value for W_A can be obtained by combining equations (3.1) and (3.2) and putting $\gamma_1 = \gamma_{SV}$, $\gamma_2 = \gamma_{LV}$ and $\gamma_{12} = \gamma_{SL}$, namely

$$W_A = \gamma_{SV} + \gamma_{LV} - \gamma_{SL} \tag{3.3}$$

W_A represents a physical bond resulting from highly localised intermolecular dispersion forces which, in the ideal situation, can give very strong adhesion between resin and carbon or glass fibres. However, as will be shown later, this strong physical bond is usually not achieved (*a*) because the fibre surface is contaminated so that the effective surface energy is much smaller than that of the base solid, (*b*) because of the presence of entrapped air and other gases at the solid surface, and (*c*) because of the occurrence of large shrinkage stresses during the curing process which leads to displacements at the surface which cannot be healed. Wetting is particularly important in composite material fabrication processes which require, for example, pick up of resin by fibre tows and the impregnation of resin into fibre bundles.

(*b*) *Interdiffusion.* It is possible to form a bond between two polymer surfaces by the diffusion of the polymer molecules on one surface into the molecular network of the other surface, as illustrated schematically in Fig. 3.3*a*. The bond strength will depend on the amount of molecular entanglement and the number of molecules involved. Interdiffusion may be promoted by the presence of solvents and plasticising agents and the amount of diffusion will depend on the molecular conformation and constituents involved and the ease of molecular motion. Interdiffusion may account in part for the bonding achieved when fibres are pre-coated with polymer before incorporating into the polymer matrix. The phenomena of interdiffusion has been called autohesion in relation to adhesives.

(*c*) *Electrostatic attraction.* Forces of attraction occur between two surfaces when one surface carries a net positive charge and the other surface a net negative charge as in the case of acid–base interactions and ionic bonding (see Fig. 3.3*b*). The strength of the interface will depend on the charge density. Although electrostatic attraction is unlikely to make a major contribution to the final bond strength of fibre–matrix composites, it could well have an important role in the way that coupling agents are laid down on the surface of glass fibres. Also the surface may exhibit anionic or cationic properties depending on the oxides in the glass and the pH of the aqueous solution used to apply the silane coupling agents. Thus, if ionic functional silanes are used, it is expected that the cationic functional groups will be attracted to an anionic surface and vice versa (Fig. 3.3*c*). It follows that by controlling pH the silane molecules can be oriented on the glass surface to obtain an optimum coupling effect which will almost certainly involve some chemical bonding (see below) since electrostatic attraction alone will not be resistant to water.

(*d*) *Chemical bonding.* This is of particular interest for fibre composite materials because it offers the main explanation for the use of coupling agents on glass fibres which is described in detail in Section 3.3 and the strength of the bond between carbon fibres and polymer matrices described in Section 3.4. A chemical bond is formed between a chemical grouping on the fibre surface and a compatible chemical group in the matrix (Fig. 3.3*d*). The strength of the bond depends on the number and type of bonds and interface failure must involve bond breakage. The processes of bond formation and breakage are in some form of thermally activated dynamic equilibrium.

Fig. 3.3. (*a*) Bond formed by molecular entanglement following interdiffusion. (*b*) Bond formed by electrostatic attraction. (*c*) Cationic groups at the end of molecules attracted to an anionic surface resulting in polymer orientation at the surface. (*d*) Chemical bond formed between groups A on one surface and groups B on the other surface. (*e*) Mechanical bond formed when a liquid polymer wets a rough solid surface.

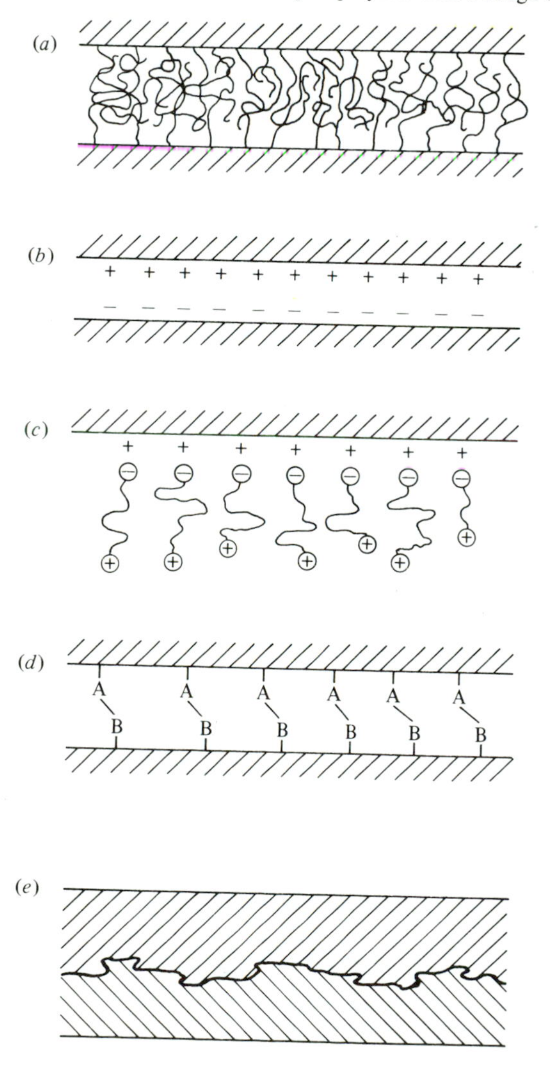

(*e*) *Mechanical adhesion.* Some bonding may occur purely by the mechanical interlocking of two surfaces as illustrated in Fig. 3.3*e*. A resin which completely wets the fibre surface follows every detail of that surface. The strength of this interface in tension is unlikely to be high unless there are a large number of re-entrant angles on the fibre surface. The strength in shear may be very significant and depends on the degree of roughness. A quite separate factor which also relates to the roughness of the fibre surface is the potential for an increased bond strength through, for example, chemical bonding because of the larger surface area which is available.

In addition to the simple geometrical aspects of mechanical adhesion, there are many internal stresses in a composite material which develop during processing operations and mechanical testing which affect the apparent strength of the fibre–matrix bond. Thus, for example, resin shrinkage during curing of thermosetting polymers and differential thermal expansion of the matrix and fibres can produce tensile, compressive and shear stresses at the interface depending on the geometry of the fibres and the component.

3.3 Glass fibre–polyester resin interface

In Section 2.2 reference was made to the application of a size to the surface of the glass fibres immediately after they have been drawn. The size fulfils many functions which include providing a chemical link between the glass and the resin. In some glass processing operations such as weaving of fibres into cloth it is appropriate to coat the fibres with a size before weaving and then to burn the size away by a process called heat cleaning and then to apply a *finish* to the fibres whose sole function is to act as a coupling agent. The thickness of the finish is less than that of the size. This section is concerned primarily with the coupling agent function of the size or finish. Many theories have been proposed to explain the role of the coupling agent; these are reviewed by Plueddemann (1974). Rather than detail all these theories the main features of the coupling action is described.

The surface of glass consists of randomly distributed groups of oxides and the composition depends on the composition of the glass (see Table 2.2). Some of the oxides, such as SiO_2, Fe_2O_3 and Al_2O_3, are non-hygroscopic and absorb water as hydroxyl groups (–M—OH, where M is Si, Fe and Al) and as molecular water which is held to the hydroxyl groups by hydrogen bonding. Other oxides are hygroscopic and when water is absorbed at the surface they become hydrated. Thus, glass picks up water very rapidly to form a well-bonded surface layer which may be many molecules thick. Long contact with water results in the solution of the hygroscopic elements leaving a porous surface made up of a network of non-hydrated oxides.

The formation of a water layer cannot be avoided in commercial processing and indeed the coupling agent is usually applied in a water-based size. The presence of water has a large effect on the wettability of the glass fibres because it leads to a large reduction in the surface energy to values in the range 10–20 mJ m^{-2} compared with over 500 mJ m^{-2} for dry virgin glass. Part of the function of the coupling agent is to increase the surface energy to ensure good wetting. Thus, the critical surface tension of a typical silane coupling agent applied to glass fibre from a water solution is 40 mN m^{-1}.

The primary function of the coupling agent is to provide a strong chemical link between the oxide groups on the fibre surface and the polymer molecules of the resin. The basic principles are illustrated in Fig. 3.4. The general chemical formula for the silance coupling agents is $R—SiX_3$. This is a multifunctional molecule which reacts at one end with the surface of the glass and at the other end with the polymer phase. The X units represent hydrolysable groups bonded to silicon (e.g. the ethoxy group $-OC_2H_5$). They are present only as intermediates since, in the aqueous size solution, they are hydrolysed to yield the corresponding silanol (Fig. 3.4*a*). The trihydroxysilanols are able to compete with water at the glass surface by hydrogen bonding with the hydroxyl groups at the surface (Fig. 3.4*b*). When the sized fibres are dried, water is removed and a reversible condensation reaction occurs between the silanol and the surface, and between adjacent silanol molecules on the surface. The result is a polysiloxane layer bonded to the glass surface (Fig. 3.4*c*). Thus the silane coated fibre presents a surface of R groups to the uncured polyester or epoxide resins. During the curing process reactive groups in the resin react with the organo-functional R groups so that they are strongly linked to the cured resin (Fig. 3.4*d*). It is essential that the R group is chosen so that it is compatible with the chemical make-up of the resin. Many different types of silane are available commercially.

The silane coating leads to a strong water resistant bond. In the absence of the silane the glass–resin interface deteriorates rapidly in the presence of water which can diffuse through the resin and attack the fibre by the hydration processes mentioned earlier. These and other changes at the fibre surface produce a large reduction in the strength of the fibres.

The relatively simple model for the effect of the silane coupling agent described above requires further development if it is to account for all the properties of the interface. In particular the interface shown in Fig.

3.4*d* will produce a strong rigid bond which is expected to fail at very low strains. Since some shrinkage during curing and differential thermal shrinkage cannot be avoided, a very rigid interface will fail long before the composite material is put into service. To overcome this difficulty, two additional properties of the interface have been introduced. Firstly, it is assumed that there is a modified zone of resin around the fibre which has mechanical properties between those of the fibre and the resin. The

Fig. 3.4. Function of coupling agent (*a*) Hydrolysis of organo-silane to corresponding silanol. (*b*) Hydrogen bonding between hydroxyl groups of silanol and glass surface. (*c*) Polysiloxane bonded to glass surface. (*d*) Organo-functional R group reacted with polymer matrix.

(*a*) $R{-}Si\,X_3 \;+\; H_2O \longrightarrow R{-}Si\,(OH)_3 \;+\; 3HX$

(*b*)

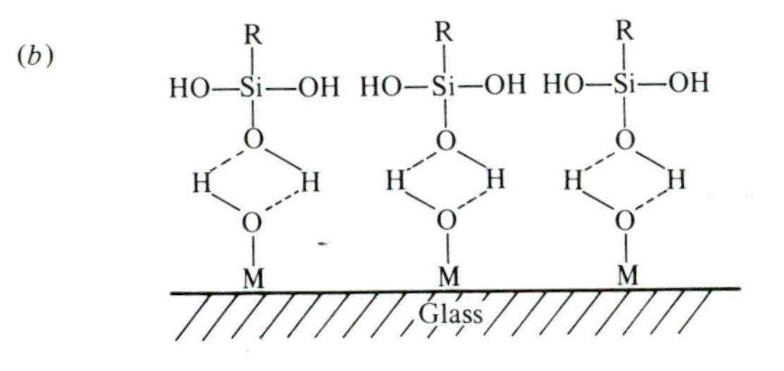

(*c*)

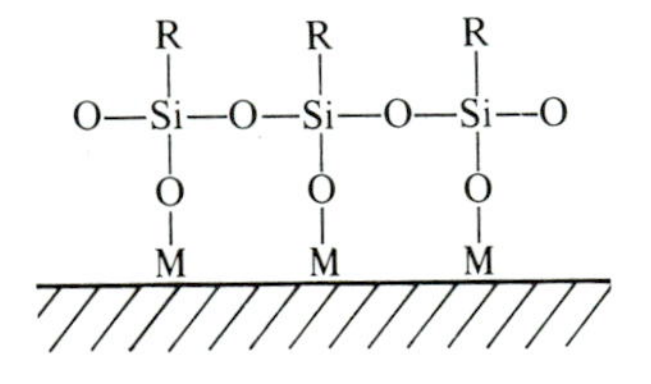

(*d*)

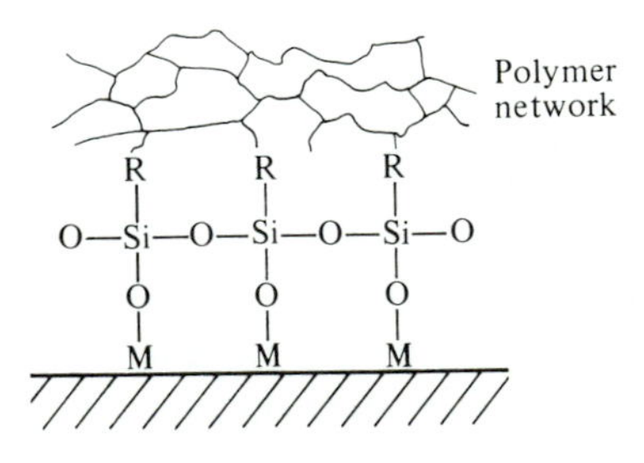

modified zone can be produced by either (i) the silane coupling agent modifying the curing behaviour of the matrix resin in the region of the fibre surface, or (ii) the coupling agent containing long R groups which, after resin cure, provide a layer of different resin between the fibre and the bulk resin. At the present time little is known about the structure, thickness or properties of these modified zones although it is widely acknowledged that such modification may have a pronounced effect on the bulk properties of composite materials. The second important feature of the interface which is a development of the simple model in Fig. 3.4*d* relates to the effect of molecular water. Plueddemann (1974) proposed that movement or displacements at the interface could relax the local stresses and maintain the chemical bond if a reversible bond breakage mechanism could occur. He proposed the mechanism shown, in its simple form, in Fig. 3.5*a*. In the presence of molecular water, which may diffuse through the resin to the interface, the covalent M—O bond hydrolyses as shown. Since this process is reversible the covalent bond can reform when the water diffuses away. Thus, in the presence of a simple shear stress parallel to the interface (Fig. 3.5*b*) the surfaces can slide past each other without permanent bond failure. Direct experimental evidence for this reversible bond process has been obtained by

Fig. 3.5. (*a*) Mechanisms of reversible bond formation associated with hydrolysis as proposed by Plueddemann (1974). (*b*) Shear displacements without permanent damage of the interface bond.

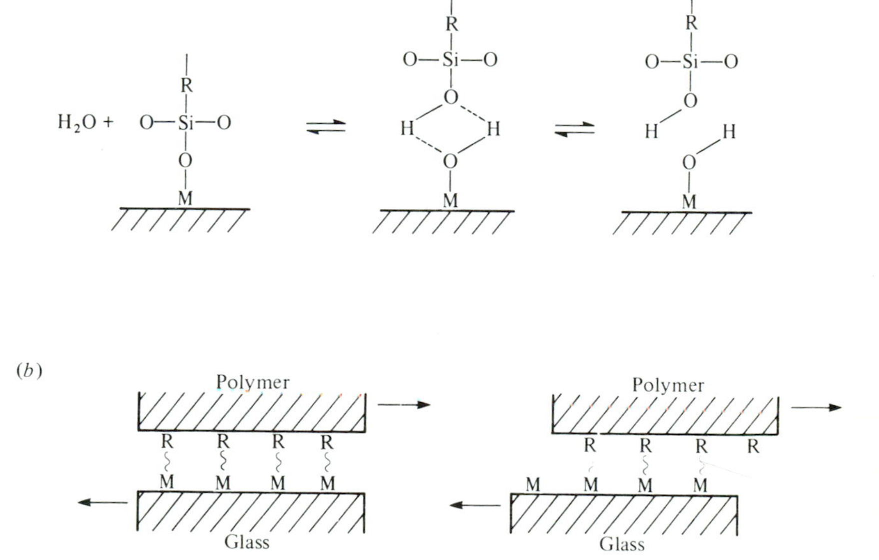

Fourier transform infra-red spectroscopy (Ishida & Koenig 1980). It is important to recognise that water has a number of other effects on composites such as plasticising the resin and causing fracture by separating as liquid water at the fibre–matrix and other interfaces (see Section 10.4).

In summary, the interface between the glass surface and the polyester or epoxy resin is a complicated one. In terms of the models for adhesion described in Section 3.2, there is likely to be a contribution from each of these mechanisms to the total bond strength although the very smooth surface of glass means that direct mechanical adhesion will be small. In some respects the 'interface' could be treated as two interfaces; the fibre–silane and the silane–resin interface. The former is invariably a chemical bond whereas the latter may be due to a combination of chemical bonding and interdiffusion if long groups are attached to the silane.

3.4 Other fibre–matrix interfaces

The preceding section has demonstrated the 'uniqueness' of the interface which is dependent on the structure and chemistry of the fibre and the resin. A similar account might be given of each fibre–matrix system although for many systems present knowledge is not so well developed and for others the information is proprietry and not available in the open literature. This section is a rather general account of some of the factors involved in other fibre–matrix interfaces.

Glass fibres

The chemical bonding mechanisms that involve silane coupling agents or other bifunctional molecules apply generally to thermosetting polymers because the organo-functional group is chemically locked into the cross-linked structure of the resin during the chemical curing reactions which change the resin from a liquid to a rigid solid. This type of chemical bonding cannot occur with glass fibres introduced into thermoplastic matrices because the molecules are already fully polymerised. However, the virgin fibre will still be susceptible to water pick up and to abrasive damage during processing so that a protective size is required. It is also important that the surface is fully wetted by the plastic and that some bonding occurs. The fibres are given a size treatment which includes a silane coupling agent and a film-forming resin. This ensures protection from water degradation and damage during injection moulding processes. It is also possible to obtain some chemical bonding if there are reactive side groups on the molecules of the thermoplastic.

The amount of reactivity varies from one thermoplastic to another. For the three thermoplastics referred to in Section 2.6 the most reactive is nylon 6.6 followed by polycarbonate and then by polypropylene. Some reactivity can be achieved by 'tailoring' the unreactive molecule so that it contains special functional groups which can bond with the coupling agent. Another approach is to include chemicals in the size which induce local chain scission of the molecules near the fibre and allow chemical reaction so that the coupling occurs directly into the molecule. Whatever method of achieving a bond is used, the effect of such bonding on the mechanical properties cannot be over-emphasised, and is particularly important in short fibre composite materials.

Carbon fibres

The bonding between carbon fibres and various polymer matrices is equally complex to that of glass fibres. A useful review of many of the important factors involved has been written by Scola (1974). The surface structure of carbon fibres is described in Section 2.1. The layer or basal planes meet the free surface at all angles between 0° and 90° and the actual structure depends on the precursor, processing route and final heat treatment temperature. Thus, for example, the high modulus PAN-based fibres have a thin skin with basal planes predominantly parallel to the surface.

Carbon is a highly active surface and readily adsorbs gases which affect the surface properties. A range of active functional groups like $-CO_2H$, $-C{-}OH$ and $-C{=}O$ can be produced on the surface by oxidative treatments such as heating in oxygen or treatment in nitric acid and sodium hypochlorite. The groups form preferentially at the edges of the basal planes and at defect sites in the basal planes. The functional groups can form chemical bonds directly with unsaturated resins and with unsaturated groups in thermoplastic resins. For some applications silane and other coatings are applied.

The reactivity of the surface is a major contributor to the strong bonding associated with carbon fibres. An additional factor is the high specific surface area due to the large amount of surface microroughness (see Fig. 2.11*a*). Thus, there are a large number of sites for chemical bonding and there is a large area of contact with the resin. It is possible to increase the apparent bond strength by increasing the specific surface area.

Fibres with a skin of basal planes aligned parallel to the surface are susceptible to cohesive failure rather than adhesive failure because of the weak bonding between the planes. It has been suggested that some of the oxidative treatments remove these surface layers so that bonding can occur onto the non-aligned layers and so avoid the weak layer plane adhesive fracture.

Kevlar 49 fibres

Relatively little has been published on the bonding of Kevlar 49 to thermosetting or thermoplastic matrices. Some surface damage can occur during processing operations such as weaving and a polyvinyl alcohol (PVA) size treatment is given to minimise this damage. It has been reported that conventional coupling agents are not particularly effective and so it is necessary to develop special systems. The fibre surface shows affinity for some epoxy resins and the fibres may be given a light pre-sizing treatment with an epoxy resin to give a better bond with other polymer matrices.

3.5 Measurement of bond strength

There is no satisfactory method of measuring the strength of the bond between the fibre and the matrix but, nevertheless, it is very important to have some measure of the strength for the evaluation of composite properties and the development of well-designed interfaces. In the first part of this section there is a simple account of the underlying problems associated with the measurement of strength and this is followed by an explanation of the distinction between adhesive and cohesive failure. Finally, some of the better established tests are described.

Consider two solids A and B bonded together and tested in uniaxial tension as illustrated in Fig. 3.6. If A and B behave as linear elastic solids and the bond strength is less than the strength of both A and B then the stress–strain curve to failure will be as shown and separation will occur at the interface at a critical stress σ_F. The work done per unit area in creating two new surfaces, W, is given by the area under the stress–strain curve

$$W = \tfrac{1}{2}\sigma_F \epsilon = \gamma_A + \gamma_B - \gamma_{AB} \tag{3.4}$$

Fig. 3.6*a* represents an ideal case since it assumes that the interface is perfect. Suppose there is a small crack length c in the interface (Fig. 3.6*b*). From the theory due to Griffith the fracture strength of the solid will be

$$\sigma_F = (\alpha E \gamma_s / c)^{\frac{1}{2}} \tag{3.5}$$

where α is a geometrical parameter which relates to the shape of the crack and test sample, E is Young's modulus and γ_s is given by

$$\gamma_s = \gamma_A + \gamma_B - \gamma_{AB} \tag{3.6}$$

The fracture strength is therefore inversely proportional to $c^{\frac{1}{2}}$ and approaches infinity at very small crack lengths. However, experience in fracture studies in a wide range of materials shows that there are always inherent flaws in the material or, alternatively, flaws are nucleated by plastic deformation which leads to a limiting fracture strength corresponding to an inherent flaw size c_0 given by

$$\sigma_F = (\alpha E\gamma_S/c_0)^{\frac{1}{2}} \tag{3.7}$$

Since solids A and B do not have the same modulus it is not strictly correct to use a single value of E in these equations but this does not affect the underlying arguments.

Fig. 3.6. Schematic representation of differences in measured bond strength due to flaws and plastic flow. It is assumed that fracture occurs at the interface. Shaded regions represent work done in fracture. (*a*) Two elastic solids with no flaws at the interface. (*b*) Two elastic solids with a flaw of length c at the interface. (*c*) One elastic and one plastic solid with no flaws at the interface.

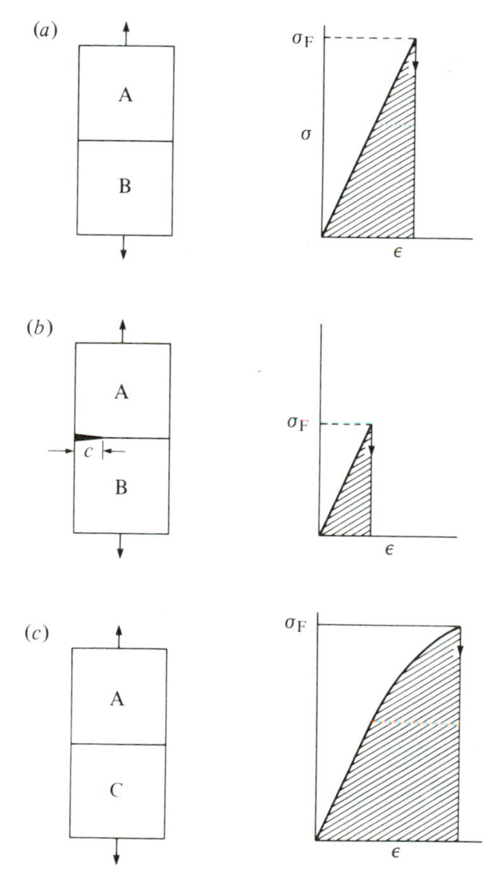

Thus the measured fracture strength of the interface depends on a parameter c_0 which is not normally amenable to direct measurement. It is also dependent on the non-linear properties of solids A and B. Consider the example in Fig. 3.6*c*. In which solid A is elastic and solid C undergoes some plastic deformation before fracture occurs. The apparent work done in creating the new surfaces i.e. the area under the stress–strain curve, includes an elastic and a plastic term i.e. $\gamma_S + \gamma_P$, where $\frac{1}{2}\gamma_P$ is the plastic work per unit area of new crack surface. The fracture strength becomes

$$\sigma_F = [\alpha E(\gamma_S + \gamma_P)/c_0]^{\frac{1}{2}} \tag{3.8}$$

In the context of the bond strength of the fibre–matrix interface, it follows that the experimentally determined values will be dependent on the stress conditions at the interface and the properties of the matrix, which in turn depend on the distribution of the fibres and other microstructural parameters.

Adhesive and cohesive failure

A further complication in evaluating experimental results is the difficulty of establishing whether or not failure has occurred at the interface. There are three possibilities which are illustrated schematically in Fig. 3.7. True adhesive failure occurs by separation at the interface whereas cohesive failure involves fracture of either the fibre or the matrix. Clearly, the occurrence of either adhesive or cohesive failure will depend on the relative strengths of the interface and the fibre or matrix.

Experimental measurements

There are two main approaches to the determinations of bond strength, one involving tests with single fibres, and the other tests on unidirectional laminae. These are discussed in detail in a review by Chamis (1974). Two single fibre tests are illustrated in Fig. 3.8. The parallel sided specimen in Fig. 3.8*a* represents a block of resin containing a short fibre along the central axis. When this specimen is compressed along the axis parallel to the fibre shear stresses are created at the ends of the fibre because of the different elastic properties of the fibre and matrix. This topic is discussed in detail in Section 5.4. It is sufficient here to record that the relation between the applied compressive stress σ_C and the interfacial shear stress τ_S is

$$\tau_S \approx 2.5\sigma_C \tag{3.9}$$

Thus, the *shear strength* of the interface bond can be obtained by measuring the value of σ_C at which debonding is first detected at the ends of the fibre. This normally involves visual observations so that the method cannot be used for opaque resins.

Fig. 3.7. Different modes of failure of a fibre embedded in a resin matrix, (*a*) adhesive failure at the interface, (*b*) cohesive failure of resin close to the interface, and (*c*) cohesive failure of fibre close to the interface.

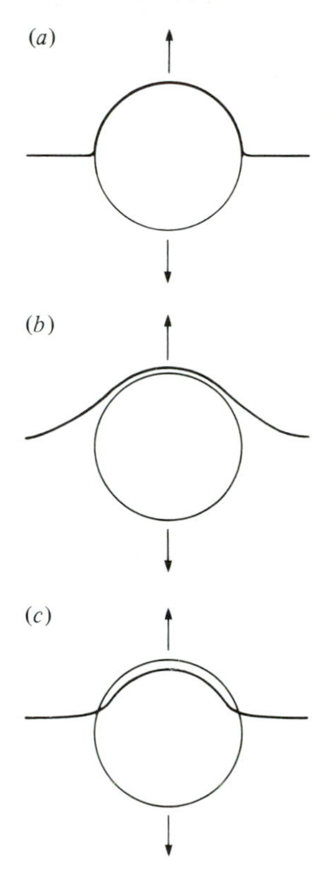

Fig. 3.8. Single fibre tests to measure (*a*) shear strength and (*b*) tensile strength of interface. (See Broutman 1969.)

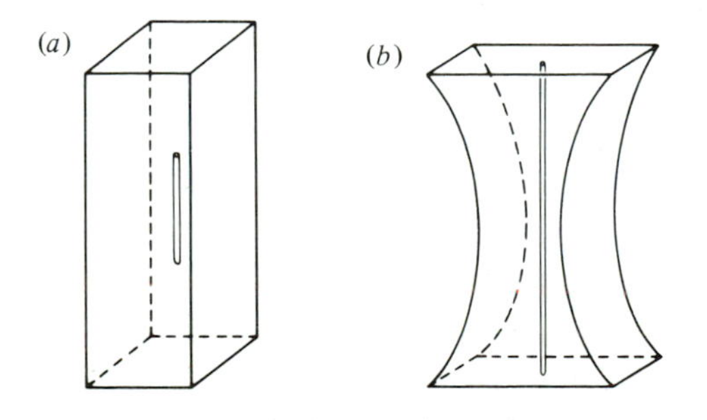

The curved neck specimen in Fig. 3.8*b* can be used to measure the *tensile strength* of the interface and is based on the same principles as the other single fibre test. When this specimen is compressed the difference in the Poisson's ratios of the resin and fibre results in a tensile stress in the centre of the neck at right angles to the fibre matrix interface which is given by:

$$\sigma_{\perp} = \frac{\sigma_C(\nu_m - \nu_f)\, E_f}{(1+\nu_m)\, E_f + (1-\nu_f-2\nu_f^2)\, E_m} \tag{3.10}$$

where ν_m and ν_f are the Poisson's ratios of the matrix and fibre respectively, σ_C is the net section compressive stress (load divided by minimum area) and E is Young's modulus. The tensile strength of the interface is obtained from the value of σ_C at which debonding occurs.

Both these tests require very precise alignment of the fibres and are inherently difficult and unreliable because of the problems associated with specimen preparation. Another type of single fibre test, which is particularly relevant to the fracture of composite materials parallel to the fibre direction, involves measuring the stress required to pull a fibre out of a block of resin or through a resin disc as described in Section 7.3. For a review of single fibre tests see Broutman (1969).

The second approach to measuring interfacial bond strength is to test unidirectional laminae in such a way that failure occurs in a shear mode parallel to the fibres or in a tensile mode normal to the fibres. Two such tests are illustrated in Fig. 3.9. The intralaminar shear test is a three-point bending test which is relatively simple and easy to perform. The shear stress $\tau_{\#}$ on the mid-plane xx' is related to the applied load P by

$$\tau_{\#} = 3P/4ab \tag{3.11}$$

where a is the thickness and b is the width of the test specimen. For the same specimen and test geometry the maximum tensile stress parallel to the fibres, which occurs on the outer fibres at point O (Fig. 3.9*a*), is given by

$$\sigma_{max} = 3PS/2a^2b \tag{3.12}$$

where S is the span or distance between the outer loading points. Since the ratio $\tau_{\#}/\sigma_{max}$ depends on the test geometry according to

$$\tau_{\#}/\sigma_{max} = a/2S \tag{3.13}$$

it follows that the likelihood of tensile or flexural failure at point O or shear failure along xx' will depend on the span to depth ratio S/a. Thus, in designing the test it is important to choose S so that failure occurs by shear. It is difficult to avoid damage by crushing at the loading points and this often complicates the interpretation of the results. Other intralaminar shear tests include torsion tests on hoop wound tubes, torsion tests on unidirectional rods and tensile tests on double notched unidirectional rods.) The transverse tensile test shown in Fig. 3.9*b* is also simple but great care is required if reliable results are to be obtained.

Both the intralaminar shear strength and the transverse tensile strength depend on fibre volume fraction as described in Section 7.4 so they cannot be regarded as giving direct values of the bond strength. All the reservations outlined earlier in this section apply. Indeed, the mode of fracture, particularly in the transverse tensile test may relate only indirectly to the bond strength. However, it is important to have some measure of the relative bond strengths of different fibre–matrix systems and these tests do have some value in this respect. Further details of the failure mechanisms can be obtained from Section 7.4 and experimental values of transverse tensile strength $\sigma_{\perp}^{*}$ and intralaminar shear strength $\tau_{\#}^{*}$ are given in Table 7.1.

An example of the effect of different surface coating treatments on the apparent interfacial shear strength measured using the single fibre compression test (Fig. 3.8*a*) and the intralaminar shear strength test (Fig. 3.9*a*) is given in Table 3.1. The glass fibres used had a diameter

Fig. 3.9. Tests on unidirectional laminae to measure (*a*) intralaminar shear strength and (*b*) transverse tensile strength.

(*a*)

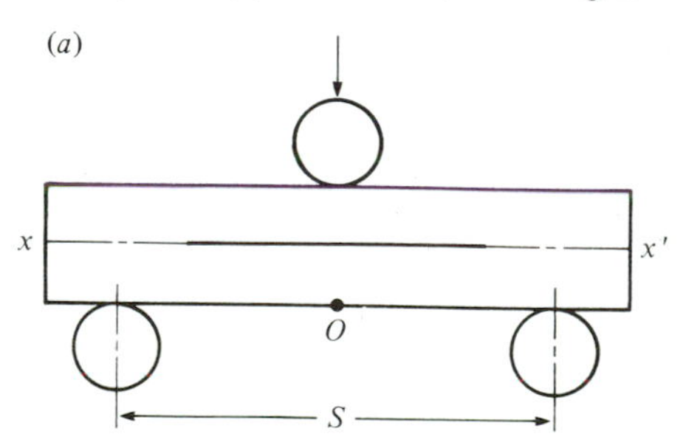

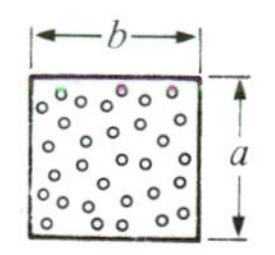

(*b*)

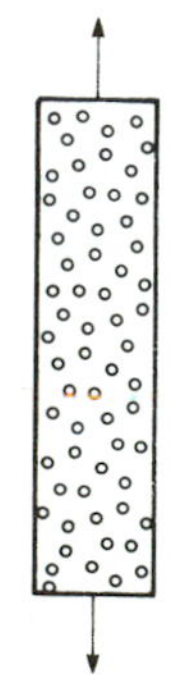

Fig. 3.10. Scanning electron micrographs of fractured specimens of injection moulded glass-filled polypropylene. (*a*), (*b*) Fibres without chemical coupling agent showing smooth fibre surfaces and holes left in the plastic matrix. (*c*), (*d*) Fibres with chemical coupling agent showing polymer adhering to the fibre surfaces. Less fibre pull-out has occurred. (Courtesy of Dow Corning Corporation.)

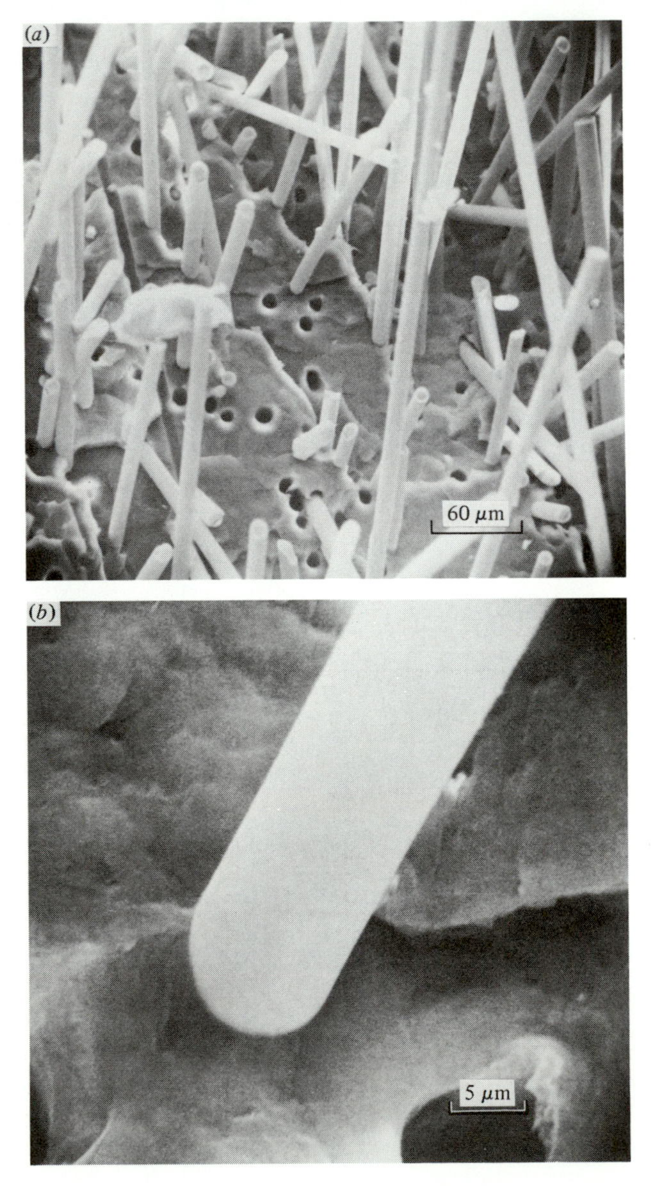

Fig. 3.10 cont.

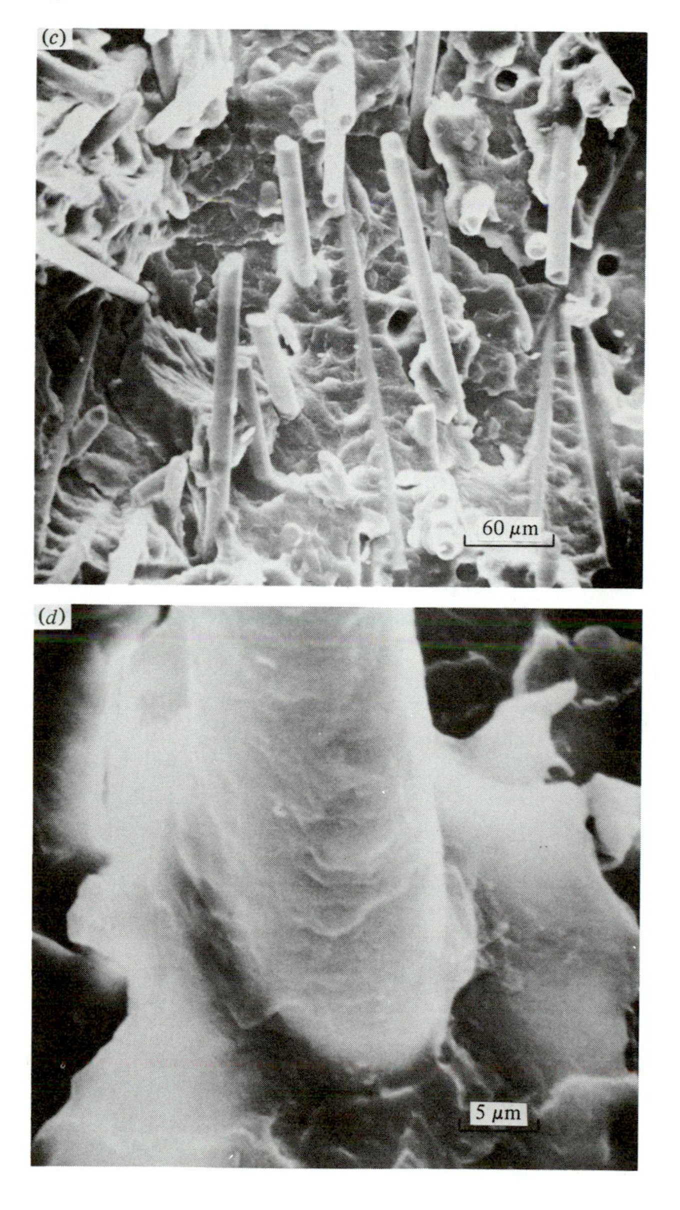

of 30 μm to facilitate the experimental work and they were heat cleaned at 773 K for 2 h before application of the surface coatings. A cold curing polyester resin was used with an MEK peroxide catalyst and a cobalt naphthenate accelerator. The silicone resin acts as an effective barrier to chemical bonding and also results in very poor wetting by the polyester resin so that no bonding occurs. It is almost impossible to

Table 3.1. *Effect of surface treatment on interfacial shear strength and intralaminar shear strength for a glass fibre–polyester resin system (after Yip & Shortall 1976)*

Surface treatment	Expected function of treatment	τ_s[a] Apparent interfacial shear strength ($MN\,m^{-2}$)	$\tau_{\#}$ Intralaminar shear strength (from short beam shear test $V_f = 0.45$) ($MN\,m^{-2}$)
Silicone resin applied in a non-aqueous solution	Unreactive towards both glass and resin	No bonding	—
0.3% silane A153 in water, pH 3.5–4.0	Reactive towards glass surface only	22.2	24.4
0.3% Morpan TBP in water	A boundary lubricant	25.2	28.6
5% polyvinyl acetate	A film forming polymer used to promote adhesion between filaments	37.1	—
Water		34.5	34.6
0.3% silane A174 in water pH 3.5–4.0	Reactive towards both the glass and polyester resin	52.3	> 41.7
0.3% silane A174 0.1% Morpan TBP in water pH 3.5–4.0	(see above)	56.7	> 49.5
0.3% silane A174 0.1% Morpan TBP 5% polyvinyl acetate	(see above)	60.5	> 39.7

[a] Obtained using equation (3.9) in the form $\tau_s = 2.0\,\sigma_c$.

make a composite material with fibres coated in this way. Some bonding is apparent with the 0.3% silane A153 when no direct chemical linking seems likely. Part of the bond strength is attributed to mechanical adhesion and part to secondary effects associated with chemical bonding. The polyvinyl acetate coating will be soluble in the polyester resin and the bond strength is similar to that obtained with the water affected surface. The strongest bond is achieved, as expected, with a silane coupling agent which reacts with both the glass and the resin to produce a true chemical bond. Resin yielding occurred before intralaminar shear fracture in the strongly bonded samples, so it was not possible to obtain a meaningful value for the shear strength of the interface.

Fracture surface observations

An indirect measure of the strength of the interface bond can sometimes be obtained from the appearance of the fracture surface as illustrated in Fig. 3.10. The materials used in this example were injection moulded thermoplastics containing short glass fibres. Fig. 3.10*a* and *b* show that when there is no coupling agent on the glass surface, the fibres are extracted from the matrix during the separation of the fracture surfaces without any residual plastic adhering to the fibres. In contrast, Fig. 3.10*c* and *d* illustrate the effect of the coupling agent in promoting fibre–matrix adhesion. Strong bonding has occurred and the surface of the glass extracted from the matrix is covered with a layer of plastic. There are corresponding changes in the appearance of the matrix fracture surfaces indicating that the degree of adhesion affects the overall fracture process.

References and further reading

Bascom, W. D. (1974) The surface chemistry of moisture-induced composite failure. *Composite Materials*, vol. 6, ed. E. P. Plueddemann, pp. 79–108. Academic Press, New York.

Broutman, L. J. (1969) Measurement of the fibre–polymer matrix interfacial strength. *Interfaces in Composites*, ASTM Special Technical Publication 452. American Society for Testing and Materials.

Chamis, C. C. (1974) Mechanics of load transfer at interface. *Composite Materials*, vol. 6, ed. E. P. Plueddemann, pp. 32–77. Academic Press, New York.

Ehrburger, P. & Donnet, J. B. (1980) Interface in composite materials. *Phil. Trans. R. Soc. Lond.* A **294**, 495–505.

Erickson, P. W. & Plueddemann, E. P. (1974) Historical background of the interface – studies and theories. *Composite Materials*, vol. 6, ed. E. P. Plueddemann, pp. 1–29. Academic Press, New York.

Hancox, N. L. & Wells, H. (1977) The effects of fibre surface coatings on the mechanical properties of CFRP. *Fibre Sci. Technol.* **10**, 9–22.

Ishida, H. & Koenig, J. L. (1980) Hydrolytic stability of silane coupling agents on E-glass fibers studied by Fourier transform infrared spectroscopy. *Proceedings of the 35th SPI/RP Annual Technology Conference*, paper 23-A. Society of the Plastics Industry, New York.

Kelly, A. (1973) *Strong Solids*. Clarendon Press, Oxford.

Plueddeman, E. P. (ed.) (1974) Mechanism of adhesion through silane coupling agents. *Composite Materials*, vol. 6, pp. 174–216. Academic Press, New York.

Scola, D. A. (1974) High-modulus fibres and the fibre–resin interface in resin composites. *Composite Materials*, vol. 6, ed. E. P. Plueddemann, pp. 217–84. Academic Press, New York.

Shortall, J. B. & Yip, H. W. C. (1976) The interfacial bond strength in glass fibre–polyester resin composite systems. I. Measurement of bond strength. *J. Adhesion* **7**, 311–32.

Wake, W. C. (1978) Theories of adhesion and uses of adhesives: a review. *Polymer* **19**, 291–308.

Yip, H. W. C. & Shortall, J. B. (1976) The interfacial bond strength in glass fibre–polyester resin composite systems. II. Effect of surface treatment. *J. Adhesion*, **8**, 155–69.

4 Geometrical aspects

4.1 Introduction

Many of the properties of fibrous composite materials are strongly dependent on microstructural parameters such as fibre diameter, fibre length, fibre length distribution, volume fraction of fibres and the alignment and packing arrangement of fibres. The effect of each parameter varies from one property to another. It is important to characterise these parameters for effective processing of composite materials and for the efficient design and manufacture of components made from composite materials.

High performance components usually consist of layers or laminae stacked up in a pre-determined arrangement to achieve optimum properties and performance. For the prediction of elastic properties each lamina may be regarded as homogeneous in the sense that the fibre arrangement and volume fraction are uniform throughout. The fibres in the laminae may be continuous or in short lengths and can be aligned in one or more directions or randomly distributed in two or three dimensions. Two simple arrangements of laminae are illustrated in Fig. 4.1. A stack of laminae is called a laminate. The flat laminate in Fig. 4.1*a* consists of identical unidirectional laminae stacked with the fibres in adjacent laminae at 90° to each other. This construction is typical of the material used for high stiffness panels in aircraft. The curved laminate in Fig. 4.1*b* is part of the wall of a cylindrical vessel and is commonly found in applications such as pressure pipes and torsion tubes. In this example the inner lamina is a layer of chopped-strand mat and the outer unidirectional laminae are arranged with the fibres oriented at $\pm 55°$ to the axis of the cylinder. In general the arrangement of laminae is more complicated because the components have to satisfy a number of different design objectives. A further complication arises in considering the overall fibre arrangement in short fibre composite materials made by compression, transfer and injection moulding techniques. The idea of discrete laminae cannot be used because there are progressive and continuous changes in fibre orientation throughout the moulded components. However, it is sometimes possible to identify regions of well-defined orientation (see Section 4.9).

This chapter is concerned primarily with the characterisation of the microstructural parameters in individual laminae and in short fibre composite materials. Brief reference is made to one topic closely related to microstructure, namely, the alignment of short fibres in flowing viscous media. This plays an important role in determining the final arrangement of fibres in components manufactured by melt processing with thermoplastic matrices, as in injection moulding, and by flow in partially cured thermosetting matrices as in compression and injection moulding of sheet moulding compounds (SMC).

4.2 Unidirectional laminae: continuous fibres

In a unidirectional lamina, Fig. 4.2*a*, all the fibres are aligned parallel to each other. In an ideal situation, and for the purposes of theoretical analysis, the fibres can be considered to be arranged on a square or hexagonal lattice as shown in Fig. 4.2*b* with each fibre having a circular cross-section and the same diameter. In practice glass and organic fibres closely approximate to a circular cross-section with a smooth surface finish, but carbon fibres, although roughly circular, may

Fig. 4.1. (*a*) Flat laminate with unidirectional laminae at 90° to each other. (*b*) Cylindrical laminate with one layer of chopped-strand mat and two unidirectional laminae.

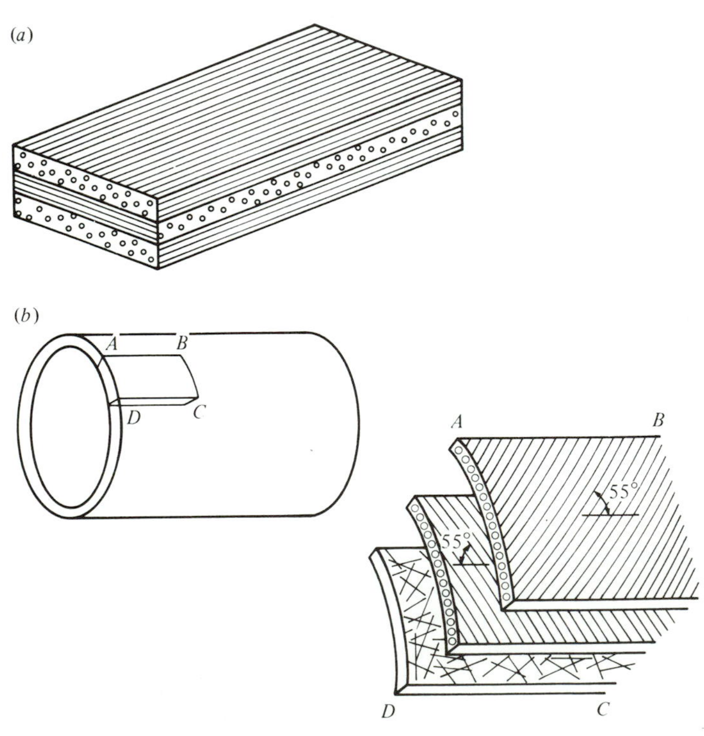

have very irregular surfaces (see Fig. 2.11*a*). There is some variation in fibre diameter for all types of material which is dependent on processing procedures.

For the ideal arrangements the volume fraction of fibres V_f is related to the fibre radius by:

$$V_f = \frac{\pi}{2\sqrt{3}}\left(\frac{r}{R}\right)^2 \quad \text{(hexagonal)} \tag{4.1}$$

$$V_f = \frac{\pi}{4}\left(\frac{r}{R}\right)^2 \quad \text{(square)} \tag{4.2}$$

where $2R$ is the centre to centre spacing of the fibres (Fig. 4.2*b*). The maximum value of V_f will occur when the fibres are touching, i.e. $r = R$. For a hexagonal array $V_{f\,max} = 0.907$ and for a square array $V_{f\,max} = 0.785$. The separation of the fibres s varies with V_f as:

$$s = 2\left[\left(\frac{\pi}{2\sqrt{3}\,V_f}\right)^{\frac{1}{2}} - 1\right]r \quad \text{(hexagonal)} \tag{4.3}$$

$$s = 2\left[\left(\frac{\pi}{4V_f}\right)^{\frac{1}{2}} - 1\right]r \quad \text{(square)} \tag{4.4}$$

Fig. 4.2. (*a*) Unidirectional lamina. (*b*) Hexagonal and square packing of unidirectional fibres.

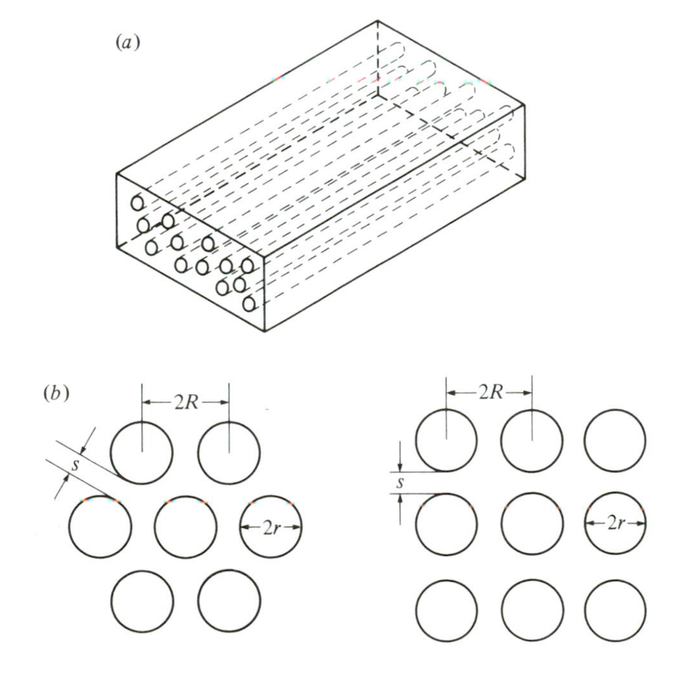

These equations are presented graphically in Fig. 4.3. Even at low V_f, for example $V_f = 0.3$, the closest distance between the fibres is less than the fibre diameter and at higher values of V_f, for example $V_f = 0.7$ the spacing becomes very small. This may have important consequences when the presence of the fibre modifies the surrounding matrix as

Fig. 4.3. Effect of V_f on spacing between fibres.

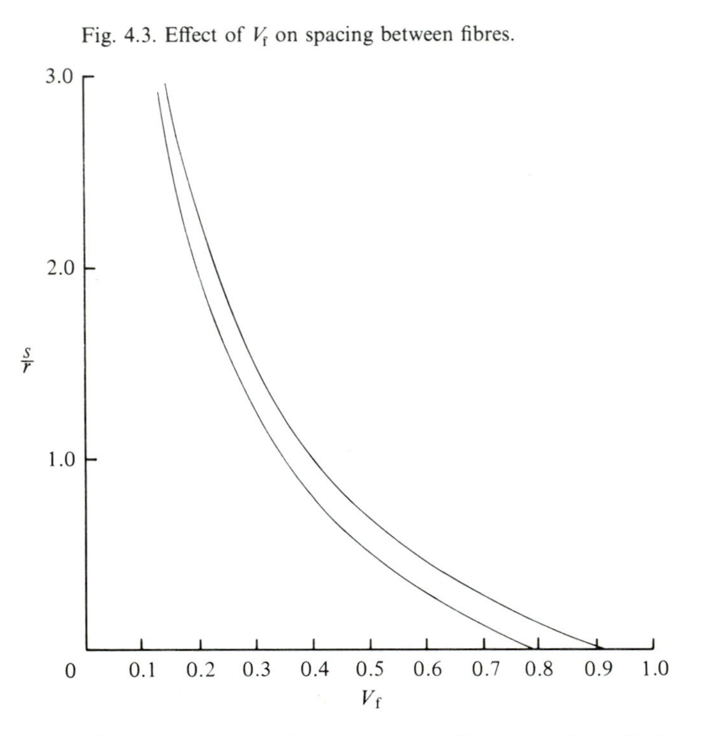

Fig. 4.4. Fibres in an hexagonal array with a layer of modified matrix around each fibre.

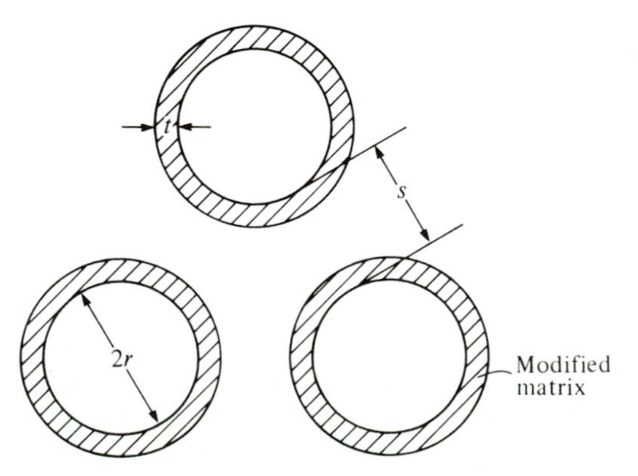

discussed in Section 3.3. Thus, for an hexagonal array of fibres surrounded by a modified matrix layer, thickness t, as shown in Fig. 4.4, the ratio of the volume fraction of modified matrix V_{mm} to the total volume fraction of matrix $V_m = 1 - V_f$ is

$$\frac{V_{mm}}{V_m} = t(2r+t)/r^2\left(\frac{1}{V_f}-1\right) \quad \text{(hexagonal)} \tag{4.5}$$

Table 4.1. *Effect of V_f and t on V_{mm}/V_m*

V_f	s(μm)	t(μm)	V_{mm}/V_m
0.4	5.06	0.1	0.027
0.4	5.06	0.2	0.054
0.4	5.06	0.5	0.140
0.4	5.06	1.0	0.293
0.6	2.29	0.1	0.061
0.6	2.29	0.2	0.122
0.6	2.29	0.5	0.315
0.6	2.29	1.0	0.660

Fig. 4.5. Photomicrograph of a section cut at right angles to fibres in unidirectional laminae of glass fibre–polyester resin.

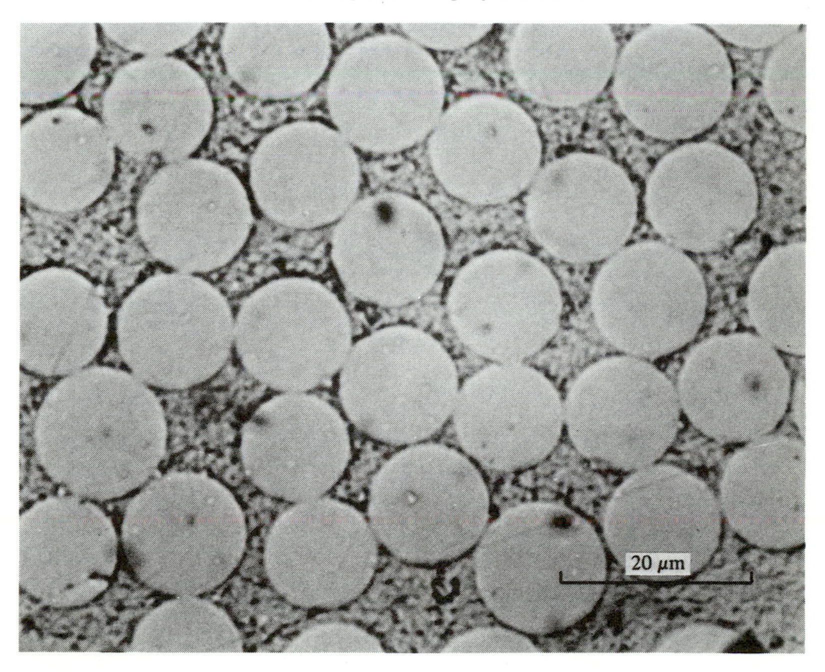

Some examples are given in Table 4.1 for an aligned composite with fibre diameter $2r = 10$ μm, $V_f = 0.4$ and 0.6. Thus, a layer of modified matrix around each fibre 0.5 μm thick makes up 31.5% of the total matrix volume in a material with $V_f = 0.6$.

Experimental studies of the distribution of fibres in unidirectional laminae show that these ideal distributions do not occur in practice except in small localised regions. An example is given in Fig. 4.5 for a fairly high V_f lamina. The section was cut normal to the fibre direction. In some regions the packing closely approximates to a hexagonal array, but resin rich regions with irregular packing occur throughout the lamina. In low V_f laminae the packing is often very irregular. Some fibre bunching and large resin rich regions may occur. Misalignment of the fibres is also more pronounced in low V_f laminae. The irregular dispersion of the fibres may have a significant effect on some properties notably the transverse strength and modulus (see Sections 5.2 and 7.4). One of the main consequences of non-regular packing is the difficulty of achieving volume fractions greater than 0.7 and this value must be regarded as the practical limit for commercial materials. It follows from Fig. 4.5 that the laminae cannot be regarded as being homogeneous from a microstructural point of view although for the prediction of laminate properties it is assumed that each lamina has a set of characteristic properties.

Another aspect of fibre packing and spacing which is important in many processing operations, particularly where the resin is introduced into a closed mould containing dry fibres, is the wetting out of the fibres by flow of the resin through the packed fibres. At high V_f and with small diameter fibres the spacing between the fibres is very small and long times and high pressures are required for complete infiltration.

4.3 Volume fraction and weight fraction

Although most calculations on composite materials are based on the volume fractions of the various constituents, it is sometimes important, particularly when calculating the density of the composite to use weight fractions. The appropriate conversion equations are:

$$V_1 = \frac{W_1/\rho_1}{W_1/\rho_1 + W_2/\rho_2 + W_3/\rho_3 \ldots} \tag{4.6}$$

and

$$W_1 = \frac{\rho_1 V_1}{\rho_1 V_1 + \rho_2 V_2 + \rho_3 V_3 \ldots} \tag{4.7}$$

where V_1, V_2 etc. are the volume fractions of the constituents, W_1, W_2 etc. are the weight fractions of the constituents, and ρ_1, ρ_2 etc. are the densities of the constituents.

Fig. 4.6. (*a*) Scanning electron micrograph of a woven roving before infiltration with resin. (*b*) Photomicrograph of a polished section through a woven roving laminate parallel to one set of fibres.

4.4 Woven roving: continuous fibres

Laminates with continuous fibres are usually manufactured by laying-up individual unidirectional laminae at pre-determined angles before final consolidation. An alternative procedure which offers some advantages in some applications is to weave the fibres into a cloth before incorporation of the resin. A typical example is shown in Fig. 4.6*a* and a polished section through a composite containing woven roving is shown in Fig. 4.6*b*. In addition to all the variables discussed in Section 4.2, a complete characterisation of woven roving composites requires details of the weave spacing, number of fibres in each roving, angle between fibres in the weave and weft directions, although this is invariably 90°, and the ratio of the number of fibres in these directions.

The woven structure leads to pockets of resin at the cross-over points and the maximum V_f for woven roving composite materials is less than for fully aligned materials.

4.5 In-plane random fibres

Laminae can be made up of long fibres, either in bundles or as individual fibres, in which the fibres lie almost entirely in one plane as illustrated schematically in Fig. 4.7. Some bending occurs at the cross-over points and the maximum V_f is much lower than in aligned laminae. Typical values for commercial laminates are in the range $V_f = 0.10$–0.30. The most important microstructural parameter is the orientation of the fibres. This can be represented by using a normalised histogram as shown in Fig. 4.8. In this example the angular intervals are 10° and the percentage of fibres in each interval has been plotted. For a completely random two-dimensional distribution the number of fibres in each interval will be independent of the angular position (histogram (*a*)). For a non-random distribution (histogram (*b*)) the preferential alignment of the fibres in some directions must be related to reference directions in the test sample, i.e. Ox and Oy.

Fig. 4.7. Schematic representation of long fibres randomly distributed in two dimensions (random in-plane).

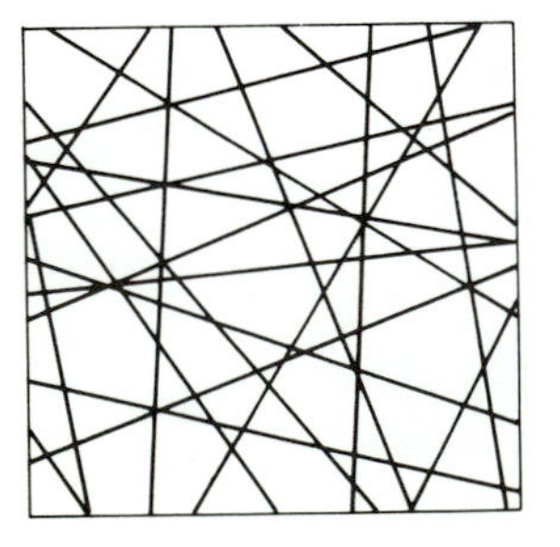

The fibre orientation distribution of long fibre planar arrays can be determined by two relatively simple techniques: firstly, for non-organic fibres, by burning off the resin and direct microscopic examination of the fibres; secondly, by microradiography which involves preparing a thin slice of the composite material parallel to the plane of the fibres and taking an X-ray transmission radiograph. The different absorption cross-sections of the fibres and resin results in the necessary contrast differences on the X-ray film. Examples of both these techniques are given in Fig. 4.9. The fibre orientation distribution can be obtained from these photographs by measurement of the orientation of each fibre or fibre bundle in a given area.

Fig. 4.8. Normalised histogram of the fibre orientation distribution of a two-dimensional array of fibres, (*a*) completely random (*b*) non-random. (The example in (*b*) is from work by Darlington, McGinley & Smith (1976) on injection moulded glass fibre–polypropylene.)

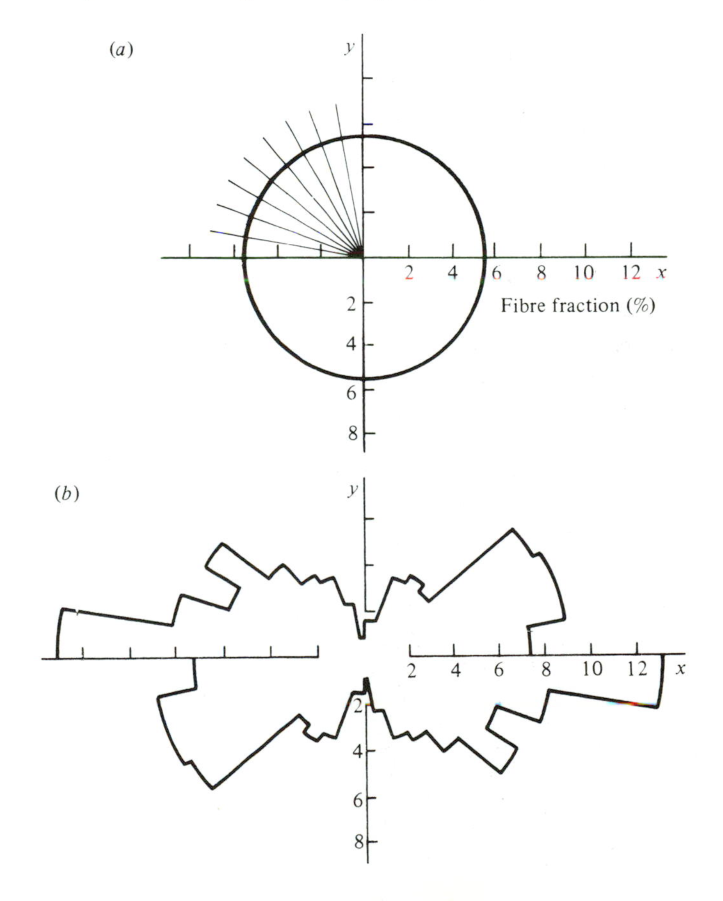

Fig. 4.9. (*a*) Scanning electron micrograph of chopped-strand mat before infiltration with resin. (*b*) Contact microradiograph of a sample with a two-dimensional array of fibres corresponding to Fig. 4.8*b*. (From Darlington, McGinley & Smith 1976).

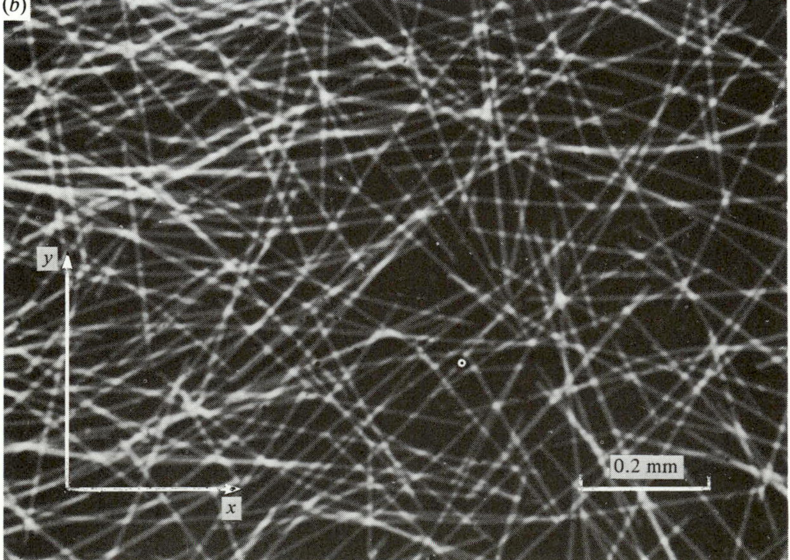

4.6 Fibre length distribution

The fibres shown in Fig. 4.9*a* were chopped to a fixed predetermined length before impregnating with resin. In making this material a thermosetting resin is introduced as a low viscosity liquid which completely wets the fibres without any fibre breakage so that the initial fibre length is retained. Incorporation of fibres into a thermoplastic matrix such as polypropylene and polycarbonate is normally achieved by blending chopped fibres with the plastic in an extruder. The process involves large shear fields which result in fibre breakage due to tensile and bending stresses. Subsequent processing such as injection moulding leads to further break-down. Since the properties of composite materials are dependent on fibre length, it is important to monitor the fibre length distribution during the processing operation to identify the factors which determine fibre fracture.

A number of methods have been used to determine the fibre length and the fibre length distribution. They can be classified broadly into indirect and direct methods. The indirect methods involve the measure-

Fig. 4.10. Optical micrograph of short fibres separated from a thermoplastic matrix after an injection moulding process. (From D. Pennington, PhD thesis, University of Liverpool 1979.)

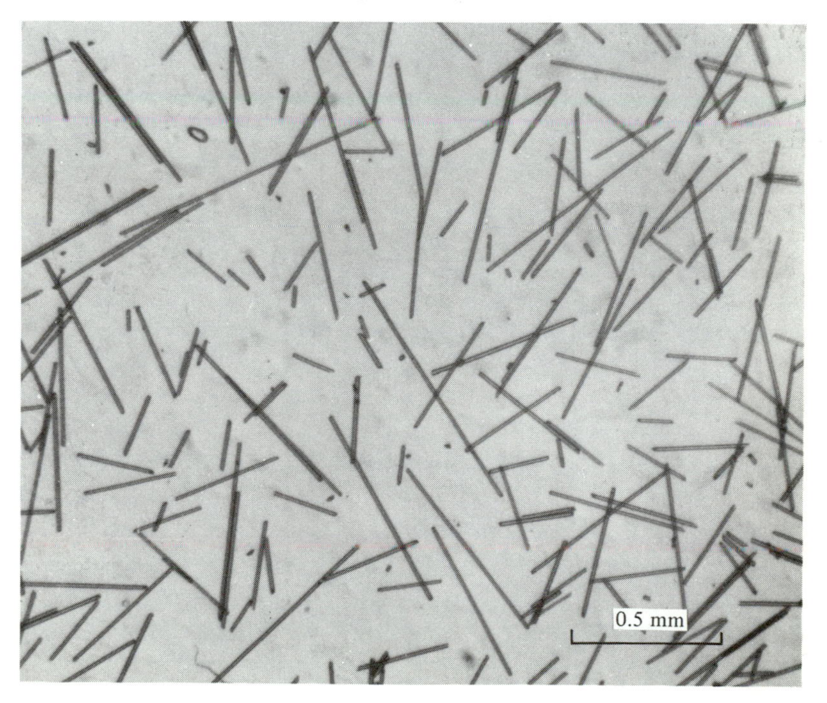

ment of some physical property of the composite, such as strength and modulus, which depends on the fibre length. This is an imprecise and unsatisfactory approach although it may have some value in quality control. In the direct methods the lengths of all the fibres in a given volume are measured. The very wide distribution of lengths and the large number of measurements which have to be made to obtain a statistically meaningful result makes this a difficult task. The first stage is the separation of the fibres from the matrix by burning off or dissolving the resin. There are two main approaches to the second stage: firstly, dispersion of the fibres in a low viscosity liquid followed by the separation of the different length fractions on a series of sieves; secondly, direct measurement of the length of each fibre by using optical microscope and photographic techniques. This method offers the most accurate and satisfactory approach but it may be tedious and time consuming. The use of automatic image analysing systems will result in more rapid evaluation.

The principles of the direct measurement method are simple and can

Fig. 4.11. Fibre length histogram of fibres extracted from a thermoset injection moulding; initial fibre length before moulding 6 mm. Detailed results from 272 measurements; smooth curve of data fitted to a lognormal distribution. (After D. Pennington, PhD thesis, University of Liverpool 1979.)

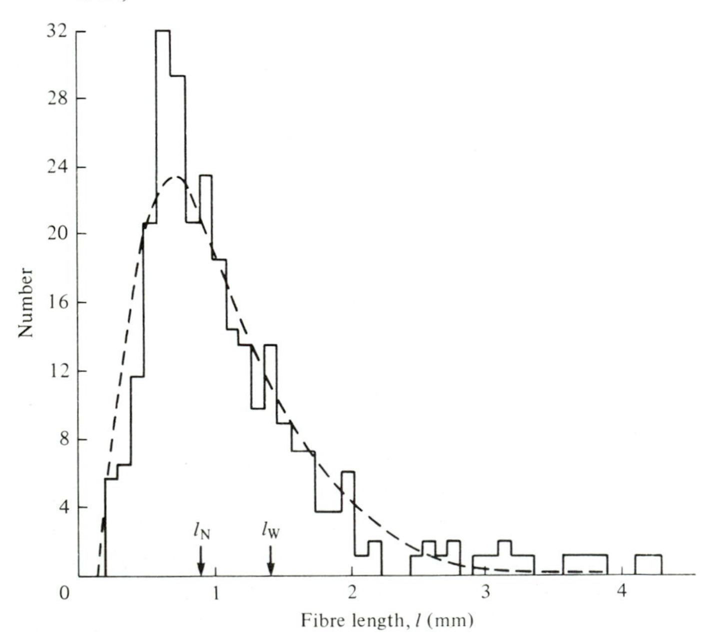

be followed by considering an example. Fig. 4.10 is a photograph of the fibres separated from a small sample of a composite material. The fibres have been dispersed. The fibre length distribution is determined by measuring the length of each fibre and plotting a histogram. A typical histogram is shown in Fig. 4.11. There is a pronounced skew distribution with a tail at the long fibre end. The sample was taken from a thermoset moulding in which the intimately mixed fibres and polymer are pressed into a mould. The initial fibre length was 6 mm and the histogram shows that none of the fibres have survived unbroken after this processing operation. The amount of fibre break-down depends on process parameters such as the design of the injection screw and shear rates in the screw, and on material parameters such as melt viscosity and fibre volume fraction.

The definition of a meaningful average fibre length is difficult but two simple averages have been used. The number average fibre length is defined as:

$$l_N = \Sigma N_i l_i / \Sigma N_i \tag{4.8}$$

where N_i is the number of fibres of length l_i. The weight average fibre length is defined as:

$$l_W = \Sigma W_i l_i / \Sigma W_i \tag{4.9}$$

where W_i is the weight of fibres of length l_i. For fibres of constant diameter this can be expressed as

$$l_W = \Sigma \alpha N_i l_i^2 / \Sigma \alpha N_i l_i = \Sigma N_i l_i^2 / \Sigma N_i l_i \tag{4.10}$$

where $\alpha = \pi r^2 \rho$ ($2r$ = diameter of fibres, ρ = density). The difference between these two averages and an indication of their physical significance is illustrated in Fig. 4.11. The number average l_N is lower than the weight average l_W and for a normal distribution coincides with the mean value of fibre length.

4.7 Fibre orientation distribution

Section 4.5 dealt with the characterisation of fibre orientation in two dimensions with particular reference to long fibres in chopped-strand mat laminae. The orientation distribution of short fibres in, for example, injection moulded products requires a three-dimensional description and is correspondingly more complicated and difficult to determine. The principle of the methods which can be used to determine the orientation of individual fibres is illustrated in Fig. 4.12. The block *ABCDEFGH* represents a thin parallel-sided slice of material which has been cut at a pre-determined position and angle with respect to the shape of the injection moulded product (i.e. the reference axes x, y, z). The orientation of the fibres crossing the thin slice can be defined in two ways:

(*a*) by angles α and β as illustrated, and

(*b*) by the shape (major and minor axes of the ellipse $2a$ and $2b$) and the orientation of the ellipse (angle α) produced by the intersection of the fibre with the parallel-sided surfaces *ABCD* and *EFGH*. It is assumed that the fibre has a circular cross-section.

For method (*a*) the angles α and β can be determined experimentally by viewing the thin section in transmission using X-ray microradiography or optical microscopy. The fibres, which are oriented in three dimensions in the thin section, appear as a two-dimensional array as illustrated in Fig. 4.13. Angle α can be measured directly from the photograph and angle β is obtained from the projected length l_p of the fibre and the thickness t of the thin section.

$$\beta = \tan^{-1}(t/l_p) \tag{4.11}$$

assuming that the fibre intersects both surfaces of the section. For

Fig. 4.12. Determination of fibre orientation in a thin section, (*a*) orientation defined by angles α and β and (*b*) orientation defined by shape and orientation of fibre cross-section.

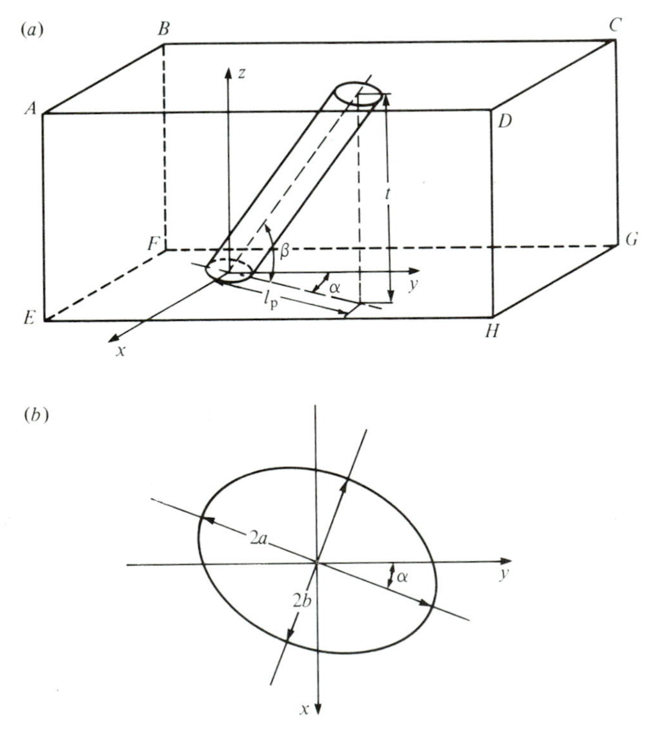

method (*b*) reference to Fig. 4.12*b* shows that the angle between the major axis of the ellipse and the reference axis y is α. Angle β can be determined from the geometry of the ellipse since

$$\beta = \sin^{-1}(b/a) \tag{4.12}$$

where $2b$ is the diameter of the fibre ($2r$) and $2a$ is the length of the major axis of the ellipse. The values of α, a and b can be determined experimentally from measurements on optical micrographs of polished sections cut from the composite material. An example is given in Fig. 4.14.

Neither of these methods fully characterises the orientation of the fibres because there are two possible positions for a fibre having angles α and β. The same projected length (method (*a*)) would be obtained from a fibre lying at an angle $\pi - \beta$ and similar arguments apply for method (*b*). It is difficult to determine which way the fibre is lying in the polished section although this can sometimes be done by visual examination of the microphotographs.

It remains to provide some graphical representation of the three-dimensional distribution corresponding to the values of α and β. Stereographic projections (see Barrett & Massalski (1966) for an account of this subject) offer one of the more straightforward methods. The values of α and β are plotted on a planar net with each point on the net representing a single fibre orientation. The population of points can then be mapped out as regions of different orientation densities in

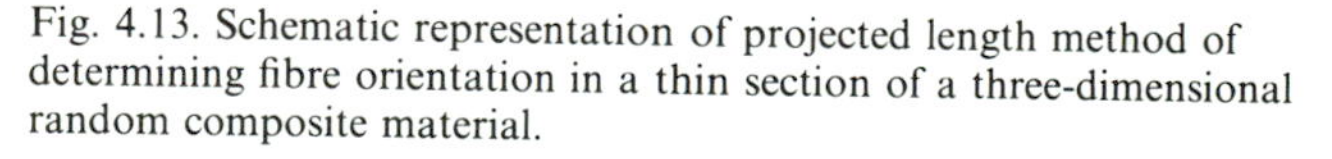

Fig. 4.13. Schematic representation of projected length method of determining fibre orientation in a thin section of a three-dimensional random composite material.

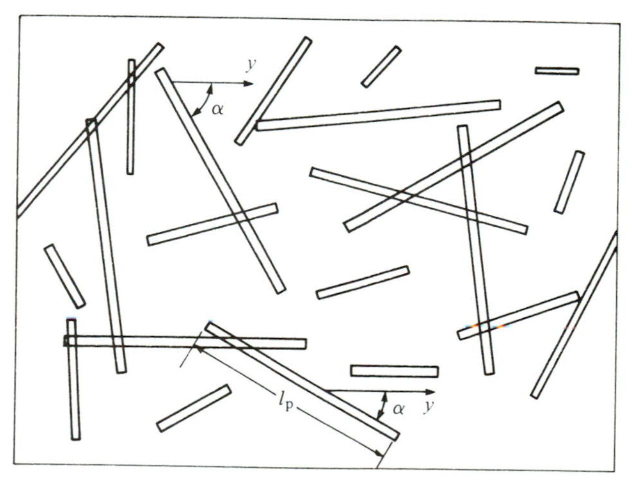

a way analogous to pole figure diagrams for grain orientation in polycrystalline metals. An example is shown in Fig. 4.15. Quantitative studies of this kind depend on obtaining a statistically representative sample and determining the orientation of every fibre in the sample. Since one gram of a typical injection moulding grade of fibre reinforced

Fig. 4.14. Optical micrograph of a polished section of a short fibre composite material (glass fibre–poly(tetramethylene terephthalate). The orientation of the fibres can be determined from the shape (a/b) and orientation of the elliptical cross-section. (From P. J. Metcalfe, unpublished work, University of Liverpool.)

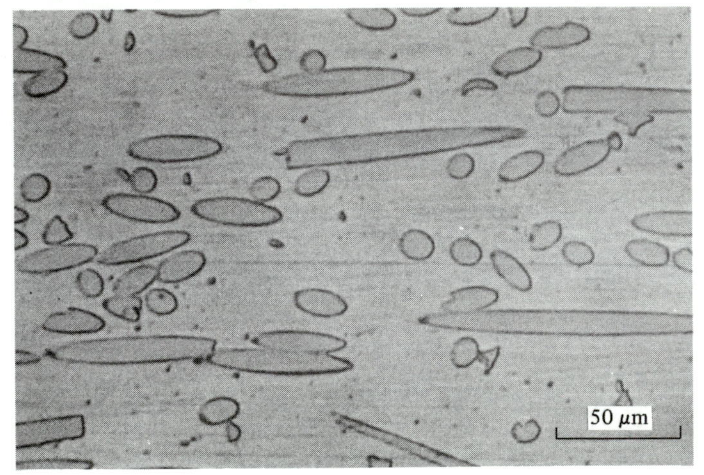

Fig. 4.15. Stereographic projection representing orientation of fibres in a thin section cut from an injection moulded flat sheet parallel to sheet. Injection moulding direction is shown and the plane of the diagram is parallel to the plane of section. Fibres are oriented preferentially in moulding direction.

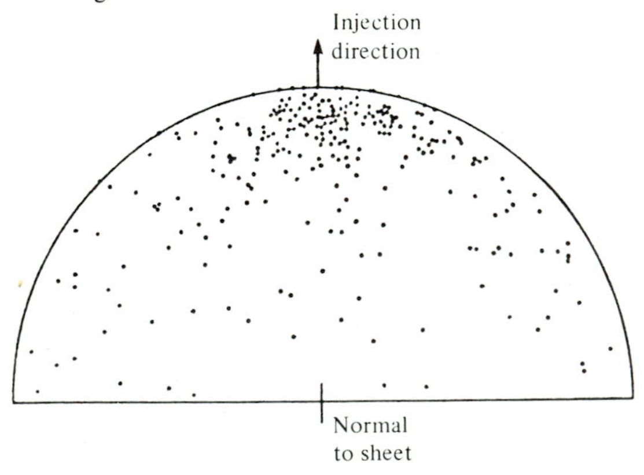

thermoplastic contains over 10^6 fibres it is difficult to obtain a sample which is large enough to be representative of the distribution but not so large that the techniques used to measure the orientation become impractical. One approach to overcome this problem (A. E. Johnson, private communication) has been to use trace fibres, for example carbon fibres in a glass fibre composite material, making the assumption that the orientation of the trace fibres is typical of the total orientation distribution.

4.8 Voids

From an evaluation of a large amount of experimental data Judd & Wright (1978) concluded that, regardless of resin type, fibre type and fibre surface treatment, the interlaminar shear strength of composite material decreases by about 7% for each 1% of voids up to a total void content of about 4%. Since other properties are also affected by the presence of voids, it is important to characterise the type of voids and void content.

Apart from large cavities due to gross defects in the manufacturing operation, there are two basic types of voids in composite materials: (i) voids along individual fibres which may be spherical or be elongated into ellipsoidal cavities parallel to the fibres; the void diameter will be

Fig. 4.16. Void between fibres in a glass fibre–polyester resin lamina.

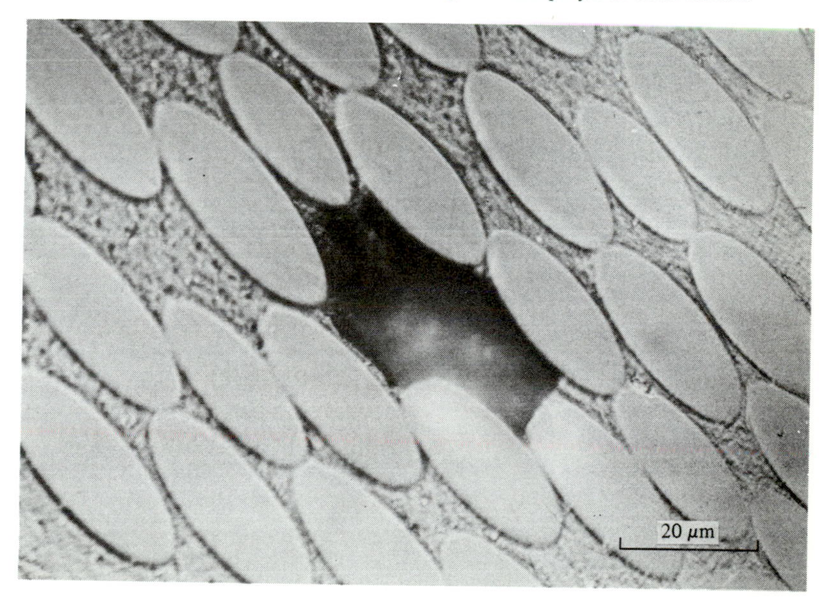

related to the fibre spacing and is typically in the range 5–20 μm; (ii) voids between laminae and in resin rich pockets. An example of the first type of void is given in Fig. 4.16.

Voids arise from two main causes: firstly, incomplete wetting out of the fibres by the resin; this results in entrapment of air and is more likely in systems where the dry fibres are closely spaced and the viscosity of the resin is high; secondly, the presence of volatiles produced during the curing cycle in thermosetting resins and during the melt processing operations in thermoplastic polymers. The volatiles may be residual solvents, products of chemical reactions or low molecular weight fractions. The void content and distribution depends on fibre volume fraction and distribution, resin properties, and processing conditions such as temperature, pressure and time.

Voids are readily detected in polished sections as shown in Fig. 4.16. One method of determining the volume fraction of voids V_v is to make a quantitative analysis, usually using point counting techniques on microphotographs. This approach allows an evaluation of the type and distribution of voids but a large number of sections must be examined to obtain a reliable value of V_v. A more simple and commonly used method is based on density measurements. The volume fraction of voids is defined as:

$$V_v = 1-(V_f+V_R) \tag{4.13}$$

or

$$V_v = 1-(W_f/\rho_f+W_R/\rho_R) \tag{4.14}$$

where W is the weight fraction and the subscript R refers to the polymer matrix. It should be noted that V_m (volume fraction of matrix) used in subsequent chapters is usually defined as $V_m = 1-V_f$ and is equal to V_R

Fig. 4.17. C-scan photograph of a multi-ply carbon fibre–epoxy resin laminate. Light regions are due to voids and delaminations in the material.

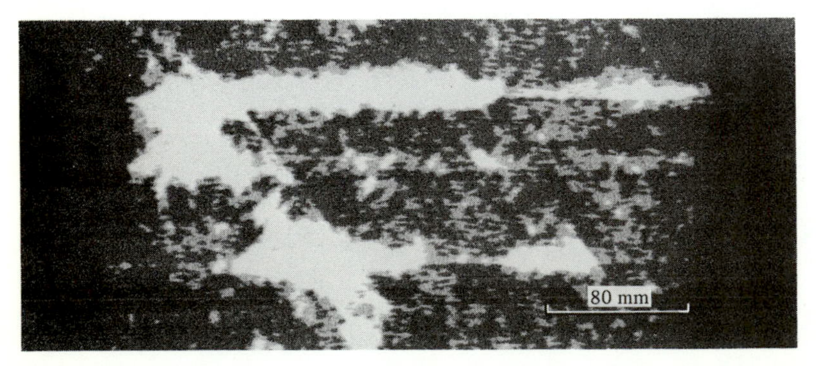

only when $V_v = 0$. In other words the void content is not taken into account separately. From equation (4.14) it follows that V_v can be determined from the weight of the fibres and the weight of the resin in a known weight of composite material. The weight of the resin can be determined from the density of the composite material and the predicted density of a void free material from a knowledge of V_f, ρ_f and ρ_R. Since V_v is determined from the difference between two relatively large values and the density measurements are not very reliable the accuracy in determining V_v is usually $\pm 0.5\%$.

Finally, an ultrasonic scanning technique (C-scan) can be used for the non-destructive examination of void distribution and delamination faults in composite materials. The test sample is scanned by an ultrasonic pulse and the attenuation in the material is measured. The information is processed to produce a two-dimensional image of the sample. The method is particularly important for the quality control of sheet materials. An example from a laminated carbon fibre–epoxy resin tube is shown in Fig. 4.17.

4.9 Fibre orientation during flow

Changes in fibre orientation occur during the processing of short fibre composite materials. The changes are related in a complex way to the geometrical properties of the fibres, the viscoelastic properties of the matrix and the change in shape of the material which is produced by the processing operation. In these operations the polymer melt undergoes both elongational or extensional flow and shear flow. An indication of the effect of these flow processes on the fibre orientation is illustrated in Fig. 4.18 for simple two-dimensional deformation. During extensional flow the fibres rotate towards the direction of extension. With large extensions a high degree of alignment can be produced. In shear flow some fibres rotate towards the direction of shear and others rotate in the opposite direction so that there is no net change in orientation. The application of shear flow to a partially oriented material results in some fibres rotating away from the principal direction of orientation. Thus the degree of preferred fibre orientation after processing is dependent on the flow field. The viscosity of the matrix affects the final orientation distribution mainly through its effect on the way in which the mould fills. This, in turn, determines the distribution of elongational and shear fields. One example will suffice.

Fig. 4.19 illustrates the mould filling process typical of an injection moulded glass fibre-filled thermoplastic. When the material is injected or extruded through the gate from the barrel of the machine into the mould cavity it experiences large elongational and compressional fields.

The material solidifies at the surface of the mould forming a skin and the mould is then filled by material which flows through the core region to the advancing front. A velocity profile is established within the core. The deformation field in the region of the solidifying skin involves a

Fig. 4.18. Schematic representation of the changes in fibre orientation occurring during flow (*a*) initial random distribution, (*b*) rotation during shear flow, (*c*) alignment during elongational flow.

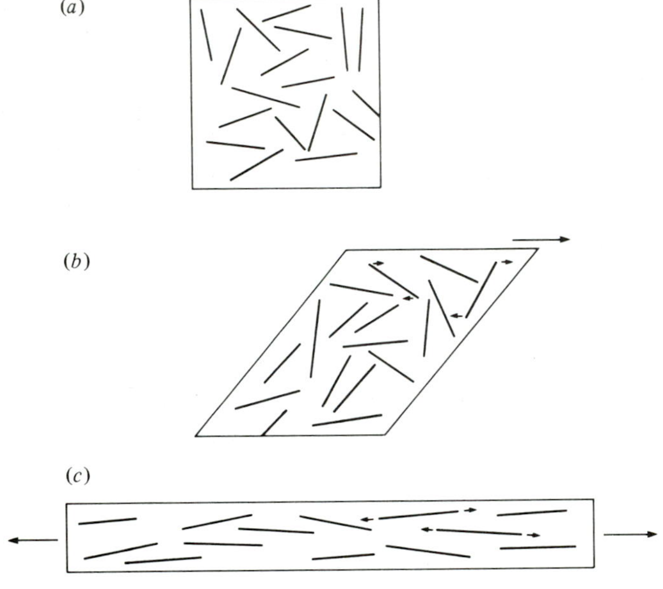

Fig. 4.19. Schematic diagram of the mould filling process showing the deformation of an initially square fluid element at successive positions of the advancing flow front. (From Folkes & Russell 1980.)

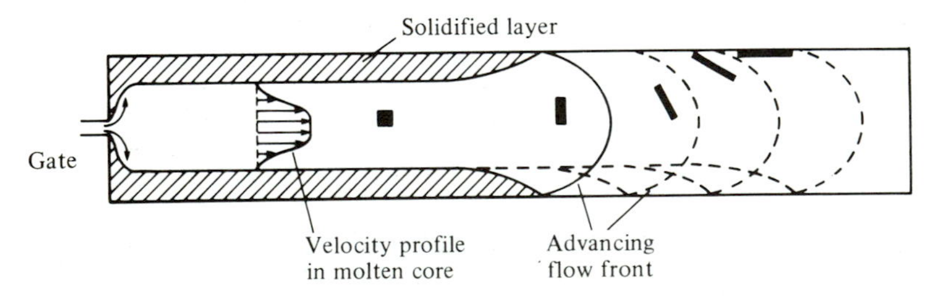

Fig. 4.20 (*See facing page.*) Contact microradiographs of (*a*) longitudinal and (*b*) transverse sections of an injection moulding. Longitudinal section corresponds to the diagram in Fig. 4.19. Material is glass-filled polypropylene, $V_f \approx 0.15$. (From Folkes & Russell 1980.)

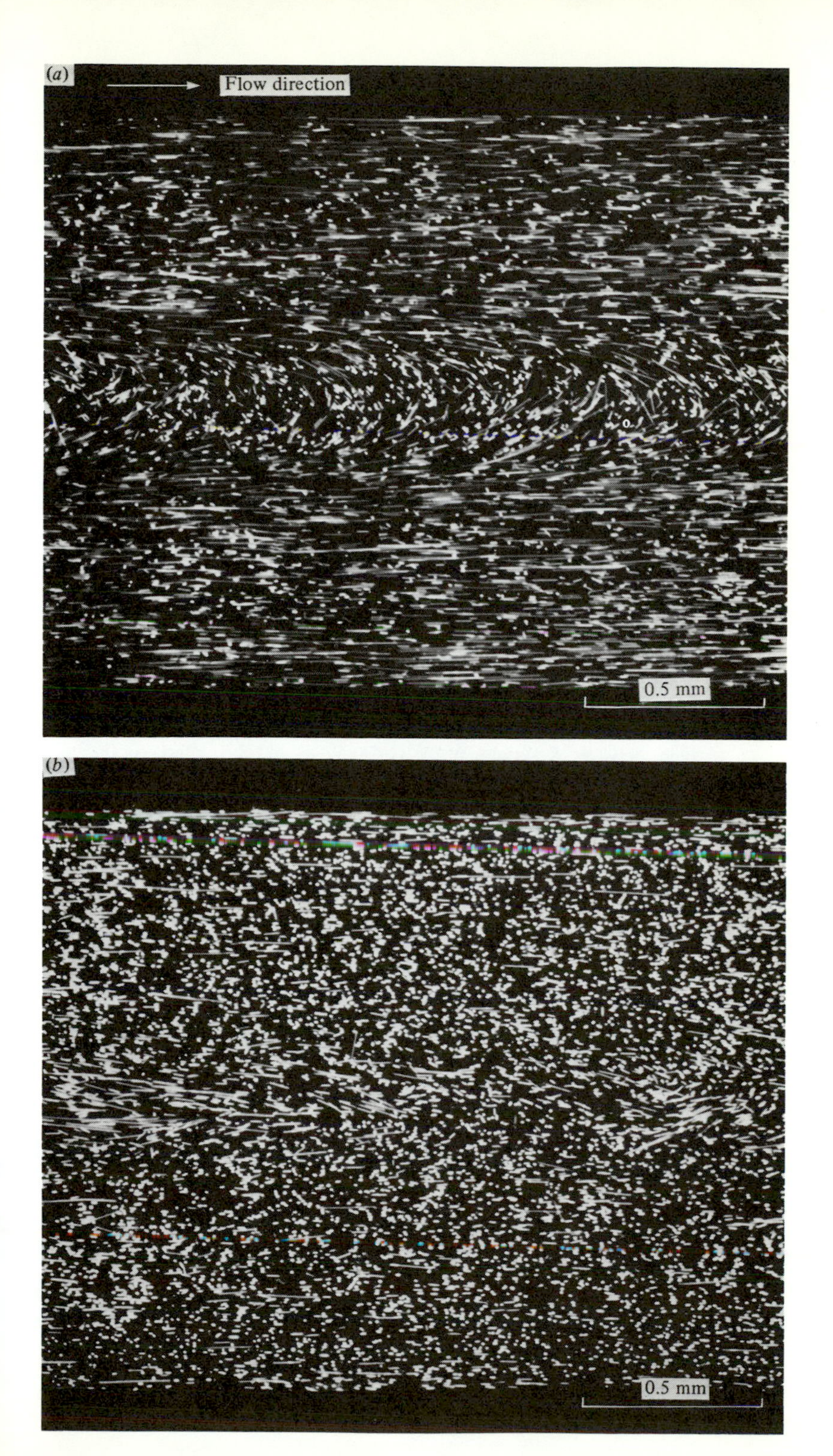
(a)
Flow direction
0.5 mm
(b)
0.5 mm

large amount of elongational flow as indicated by the change in shape of the initially square fluid element. Solidification of the core occurs, after complete mould filling, under completely different flow fields than the skin.

Although in practice the flow is much more complicated, this simple example accounts for the large difference in fibre orientation evident in the contact microradiographs in Fig. 4.20. There is a pronounced preferred orientation of the fibres parallel to the flow direction in the outer layers of the moulding and a more random distribution in the core. The extent of orientation and the type of layer structure which develops depends on process conditions such as injection temperature, injection pressure and mould temperature and also on factors such as fibre length, V_f and the dimensions and shape of the mould and gating system.

References and further reading

Barrett, C. S. & Massalski, T. B. (1966) *Structure of Metals*. McGraw-Hill, New York.

Darlington, M. W., McGinley, P. L. & Smith, G. R. (1976) Structure and anisotropy of stiffness in glass fibre-reinforced thermoplastics. *J. Mater. Sci.* **11**, 877–86.

Folkes, M. J. & Russell, D. A. M. (1980) Orientation effects during the flow of short-fibre reinforced thermoplastics. *Polymer* **21**, 1252–8.

Judd, N. C. W. & Wright, W. W. (1978) Voids and their effects on the mechanical properties of composites – an appraisal. *SAMPE Journal* Jan./Feb., 10–14.

Kelly, A. (1971) Microstructural parameters of an aligned fibrous composite. *The properties of fibre composites. Conference Proceedings of the National Physical Laboratory*, pp. 5–14. IPC Science and Technology Press, Guildford.

5 Elastic properties

5.1 Introduction

In Section 3.1 there is a simple example to illustrate the importance of the fibre–matrix interface in transferring the stress from the matrix to the fibre. Using the assumptions listed in Section 3.1 about the properties of the interface, it is possible to calculate the distribution of stress and strain in a composite material in terms of the geometry, distribution and volume fraction of the fibres and the elastic properties of the fibres and the matrix. Some of these calculations are very complex and, indeed, some problems remain to be solved. From the distribution of stress and strain the elastic properties of the composite material can be calculated. Because of the complexity of the calculations it is often necessary to use empirically determined relations.

In the first part of this chapter the relation between the elastic properties of a unidirectional lamina and the properties of the constituents are determined and compared with experimental results. This is followed by a description of the corresponding relationships for random long fibre laminae. In both the unidirectional and random long fibre laminae effects associated with fibre ends can be ignored, except when considering fracture processes. This is not true for short fibre composite materials and a section is devoted to the prediction of the stress and strain distribution in the fibre and the surrounding matrix associated with the ends of fibres. This leads on to the prediction of the elastic properties of short fibre composite materials although, here again, it is necessary to use the empirical approach except for very simple fibre arrangements.

5.2 Elastic properties of unidirectional laminae

When a tensile or compressive load is applied parallel to the fibres in a unidirectional lamina (Fig. 5.1), the strain ϵ_1 in the matrix will be the same as the strain in the fibre if the bond between the fibre and matrix is perfect. If both fibres and matrix behave elastically then the corresponding stresses are given approximately by:

$$\sigma_f = E_f \epsilon_1, \quad \sigma_m = E_m \epsilon_1 \tag{5.1}$$

It follows that if $E_f > E_m$ the stress in the fibres is greater than in the matrix. This is, of course, the underlying basis of fibre reinforcement since the fibres bear the major part of the applied load P. For a composite material with a total cross-sectional area A, the average stress is given by

$$P = \sigma_1 A \tag{5.2}$$

Since $$P = P_f + P_m \tag{5.3}$$

and $$P_f = \sigma_f A_f, \quad P_m = \sigma_m A_m \tag{5.4}$$

$$P = \sigma_f A_f + \sigma_m A_m \tag{5.5}$$

where A_f and A_m are the cross-sectional areas of the fibre and matrix respectively and P_f and P_m are the corresponding loads. Substituting equations (5.1) and the relation

$$\sigma_1 = E_1 \epsilon_1 \tag{5.6}$$

into these equations gives

$$E_1 = E_f A_f/A + E_m A_m/A \tag{5.7}$$

Now $$V_f = A_f/A \quad \text{and} \quad V_m = A_m/A \tag{5.8}$$

Fig. 5.1. (*a*) Unidirectional lamina showing directions 1, 2 and 3. (*b*) Shape change of a unidirectional lamina due to a tensile load in the 1-direction when there is a perfect fibre–matrix bond.

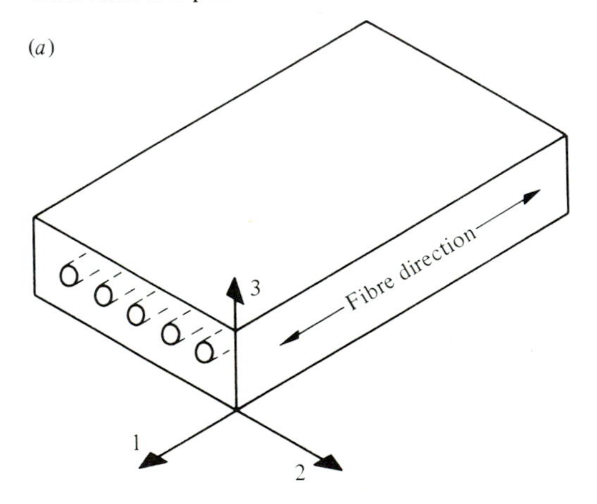

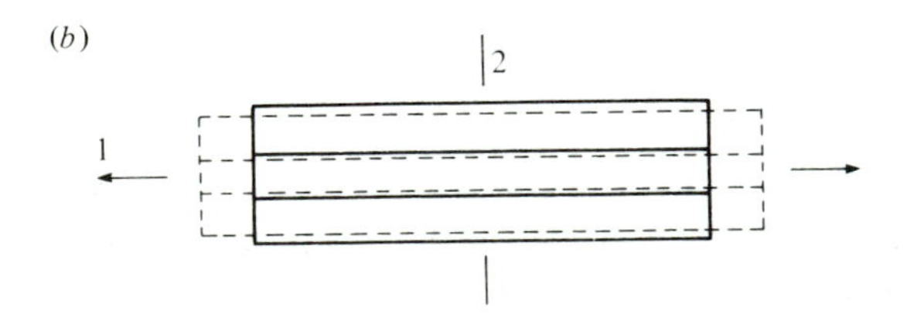

Therefore,

$$E_1 = E_f V_f + E_m V_m \tag{5.9}$$

and since

$$V_m = 1 - V_f \tag{5.10}$$

$$\dagger\ E_1 \text{ or } E_{\parallel} = E_f V_f + E_m(1 - V_f) \tag{5.11}$$

This is often referred to as the '*rule of mixtures*' equation. The analysis is based on the assumption that equation (5.1) is valid. This is not strictly true since different Poisson contractions (i.e. $\nu_m \neq \nu_f$) will result in additional stresses which have not been considered here. However,

Fig. 5.2. Elastic moduli measured parallel to fibres of undirectional laminae of glass fibres and polyester resin with different V_f. (From Brintrup Dr-Ing thesis 1975, Technischen Hochschule, Aachen.)

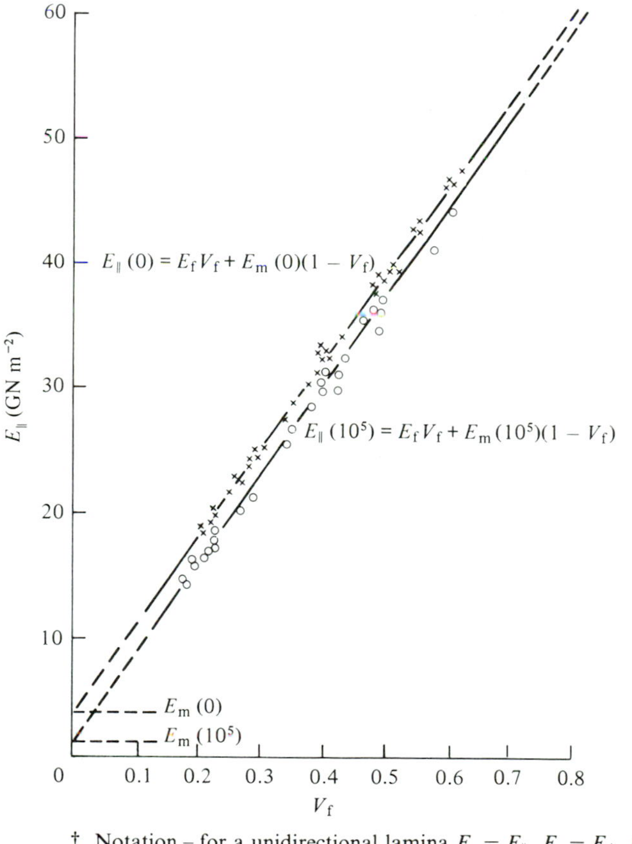

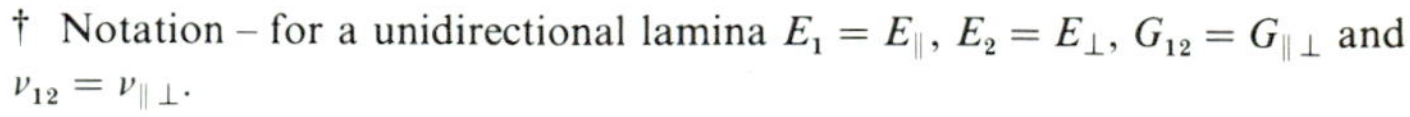

† Notation – for a unidirectional lamina $E_1 = E_{\parallel}$, $E_2 = E_{\perp}$, $G_{12} = G_{\parallel\perp}$ and $\nu_{12} = \nu_{\parallel\perp}$.

the error in $E_{\parallel}$ is likely to be less than 1 or 2% and experimental verification for equation (5.11) has been obtained for many fibre–resin systems. An example is given in Fig. 5.2 for a glass fibre–polyester resin system. The majority of the data is for V_f between 0.2 and 0.6 but in agreement with equation (5.11) the data can be extrapolated to E_m at $V_f = 0$ and to E_f at $V_f = 1$. Two sets of results are given in Fig. 5.2. $E_{\parallel}(0)$ is the value of E_1 determined from short term tests corresponding to $E_m = 3.8$ GN m^{-2}, and $E_{\parallel}(10^5)$ is the 'relaxed' modulus $\sigma/\epsilon_1(10^5)$ measured after 10^5 h with a corresponding $E_m(10^5) = 1.3$ GN m^{-2}. This illustrates the small effect that matrix creep relaxation and E_m has on $E_{\parallel}$.

The same kind of 'mechanics of materials' approach can be used to predict the transverse modulus of a unidirectional lamina $E_2 = E_{\perp}$. The simplest model of the composite material is represented in Fig. 5.3 as a slice cut parallel to the fibres. The applied load transverse to the fibres acts equally on the fibre and the matrix and the assumption is made that $\sigma_f = \sigma_m$. The corresponding strains are

$$\epsilon_f = \sigma_2/E_f, \quad \epsilon_m = \sigma_2/E_m \tag{5.12}$$

Thus, the strain ϵ_2 is given by

$$\epsilon_2 = V_f\,\epsilon_f + V_m\,\epsilon_m \tag{5.13}$$

and by substitution of equation (5.12)

$$\epsilon_2 = V_f\,\sigma_2/E_f + V_m\,\sigma_2/E_m \tag{5.14}$$

Substituting

$$\sigma_2 = E_2\,\epsilon_2 \tag{5.15}$$

in equation (5.14) and re-arranging gives,

$$E_{\perp} = \frac{E_f\,E_m}{E_f(1-V_f)+E_m\,V_f} \tag{5.16}$$

Fig. 5.3. Shape change of a unidirectional lamina due to a tensile load in the 2-direction when there is a perfect fibre–matrix bond.

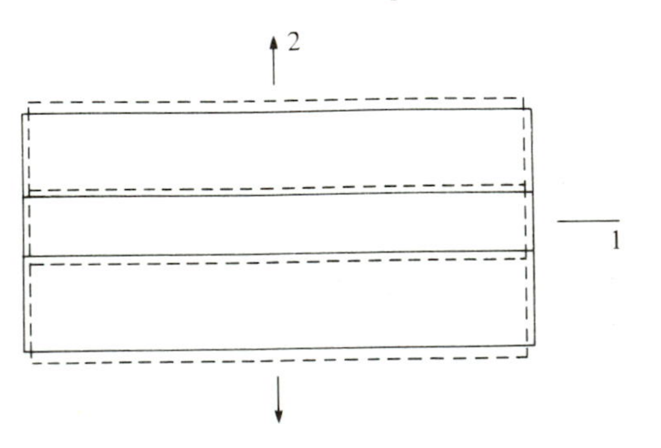

The general form of this expression is in reasonable agreement with the experimental results in Fig. 5.4 for the variation of $E_\perp$ with V_f for glass fibre–polyester resin composite materials. However, the equation is not a good fit to the actual results and there is a considerable amount of scatter. Other equations of the same form as equation (5.16) have been proposed to take account of Poisson contraction effects, which are illustrated in Fig. 5.4, and give a better fit to experimental results for some values of V_f. Thus, for example, the expression

$$E_\perp = \frac{E_{m'} E_f}{E_f(1-V_f)+V_f E_{m'}} \tag{5.17}$$

where

$$E_{m'} = \frac{E_m}{1-\nu_m^2} \tag{5.18}$$

Fig. 5.4. Elastic moduli measured transverse to fibres of unidirectional laminae of glass fibres and polyester resin with different V_f. (From Brintrup Dr-Ing thesis 1975, Technischen Hochschule, Aachen.)

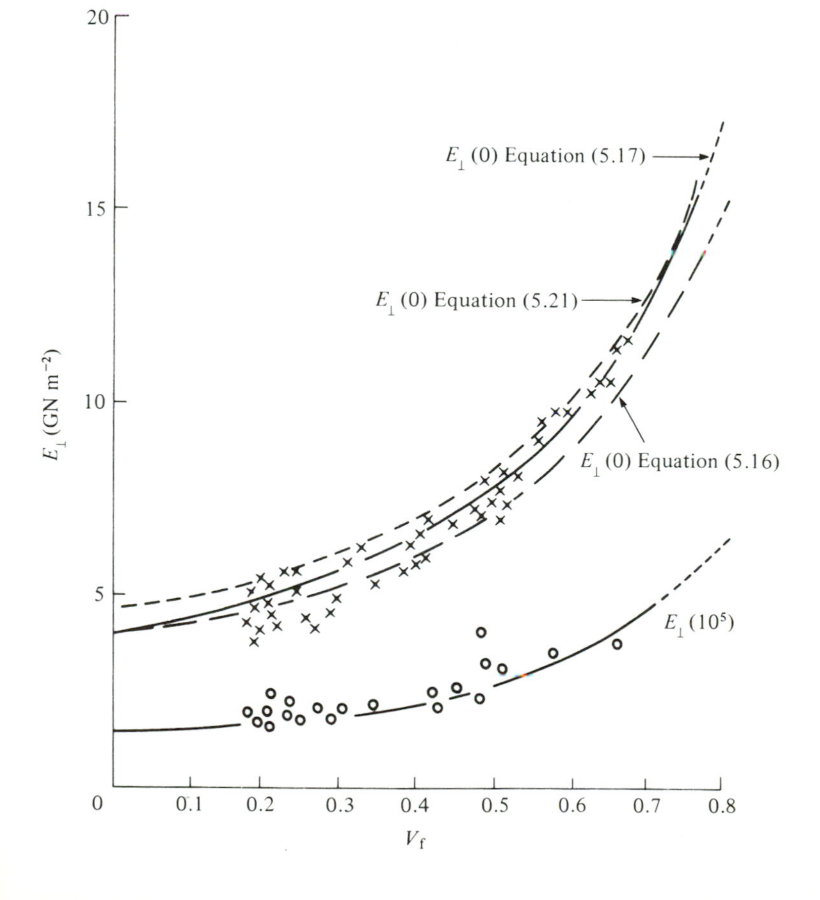

gives a curve (Fig. 5.4) which is closer to the experimental results at high values of V_f. The results in Fig. 5.4 also illustrate the pronounced effect of resin properties on the transverse modulus and, in particular, the influence of resin creep. Thus, there is a large decrease in 'relaxed' modulus with time which, in contrast to $E_{\parallel}$, is more pronounced at high V_f.

Equations (5.16) to (5.18) are essential for effective design of laminated structures but they provide little insight into a physical understanding of the stress and strain distribution around fibres which is required for an interpretation of transverse fracture (see Sections 3.5 and 7.4). The assumption $\sigma_f = \sigma_m$, and the implicit assumption, which was used to derive equation (5.16), that within each phase the strain is uniform, is unrealistic because the fibres cannot be represented as sheets as in Fig. 5.3. This can be demonstrated by reference to Fig. 5.5 which shows an idealised hexagonal array of fibres in a composite material subjected to a uniform externally applied strain. The majority of the strain in a slice XX' which corresponds to the slice represented in Fig. 5.5, is taken up by the resin since $E_f \gg E_m$. The strain in slice YY' which passes completely through the resin will be far more uniform and the average strain will be much smaller than in the resin in slice XX'. In other words there is a strain magnification in the resin between the fibres. The different strains in different parts of the resin lead to additional stresses and consequently a non-uniform stress distribution.

Elasticity theory and finite element analysis have been used to predict $E_{\perp}$ and other moduli using more realistic assumptions. A simplification of some of these solutions has been developed by Halpin & Tsai (1967) and their equations, which are set out below, are useful for the prediction of the properties of composite materials. They are more

Fig. 5.5. Schematic representation of strain magnification in a unidirectional lamina subjected to a transverse load.

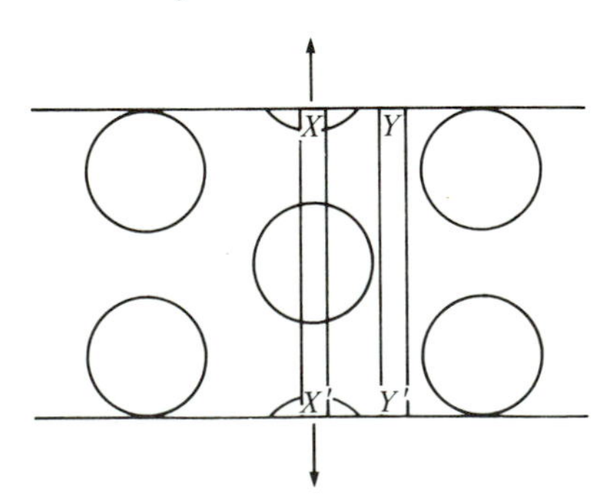

generally applicable than equations such as (5.17). The Halpin–Tsai equations are:

$$E_{\parallel} = E_f V_f + E_m(1 - V_f) \quad (5.19)$$

$$\nu_{\parallel\perp} = \nu_f V_f + \nu_m(1 - V_f) \quad (5.20)$$

and $$M_f/M_m = (1 + \xi\eta V_f)/(1 - \eta V_f) \quad (5.21)$$

where $$\eta = \frac{(M_f/M_m) - 1}{(M_f/M_m) + \xi} \quad (5.22)$$

in which M is the composite modulus $E_{\perp}$, $G_{\parallel\perp}$ or ν_{23}, M_f the corresponding fibre modulus E_f, G_f or ν_f, and M_m the corresponding matrix modulus E_m, G_m or ν_m. ξ depends on various characteristics of the reinforcing phase such as the shape and aspect ratio of the fibres, packing geometry and regularity and also on loading conditions. It is necessary to determine ξ empirically by fitting the curves to experimental results. As an example, the Halpin–Tsai equation for $E_{\perp}$ has been fitted to the results in Fig. 5.4 using $\xi = 0.2$.

Kies (1962) made one of the earliest quantitative estimates of the non-uniform distribution of strain in the matrix between the fibres using the simple model illustrated in Fig. 5.6. When a square array is subjected to a simple tensile strain $\bar{\epsilon}_x$ the strain magnification in the resin along the line AB according to Kies is

$$\epsilon_x/\bar{\epsilon}_x = 2 + \frac{s}{r}\bigg/\left[\frac{s}{r} + 2\left(\frac{E_m}{E_f}\right)\right] \quad (5.23)$$

where s and r are as defined in Fig. 5.6. Equations (4.2) and (5.23) can

Fig. 5.6. Square array model used by Kies (1962) to calculate strain magnification.

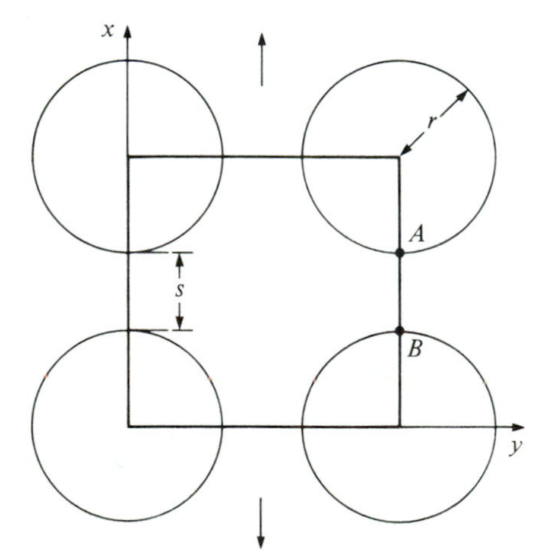

Fig. 5.7. Dependence of strain magnification on V_f according to equation (5.23) for a glass fibre–polyester resin.

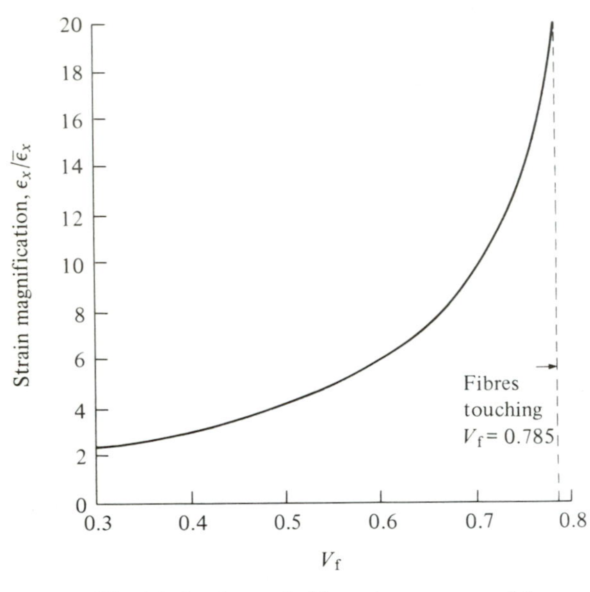

Fig. 5.8. Isochromatic fringes in a macromodel composite material loaded in transverse tension. From Puck (1967).

be combined to calculate the variation of the strain magnification with V_f. A typical set of results are shown in Fig. 5.7 which correspond to a glass fibre–polyester resin system ($E_f/E_m = 20$). Very large strain magnifications occur at high volume fractions. The strain magnification depends on the loading conditions and varies throughout the matrix.

Another approach to understanding the non-uniform distribution of stress and strain in a lamina subjected to transverse loads is the use of photoelastic techniques. The example in Fig. 5.8 shows the isochromatic fringes produced in a macromodel of a composite material in which rods representing the fibres are embedded in a resin. The fringes are seen when the model is stressed in the transverse direction and viewed in polarised light. Each fringe is a contour of the positions in the resin which have the same principal stress difference. It is possible to determine the stress and strain pattern from the position and 'order' of the fringes; see, for example, Marloff & Daniel (1969). It is sufficient to note that the fringe pattern is complex and indicates that the magnitude and direction of the stresses varies throughout the matrix and depends, for a given volume fraction, on the fibre arrangement (i.e. square or hexagonal) and on the degree of regularity. Biaxial and triaxial stresses are also produced in the matrix and their magnitude depends on the ratio of the moduli of the fibres and matrix.

From the simple approach used in determining equation (5.23) it is evident that in a unidirectional lamina with the fibres randomly distributed as shown in Fig. 4.5 the strain magnifications will vary widely from one part of the lamina to another. This may have a profound effect on the fracture processes.

The Halpin–Tsai equations (5.19) to (5.22) include expressions for the shear modulus and Poisson's ratio. Experimental verifications similar to those shown in Figs. 5.2 and 5.4 for $E_{\parallel}$ and $E_{\perp}$ have been published along with other semi-empirical equations.

5.3 Elastic properties of in-plane random long fibre laminae

A lamina made by impregnating resin into a long fibre chopped-strand mat has elastic properties which are macroscopically isotropic in the plane of the lamina provided that the fibres are randomly distributed, i.e. there is a uniform probability distribution over the entire range of angles from $-\frac{1}{2}\pi$ to $+\frac{1}{2}\pi$. For long fibres the effect of fibre ends can be neglected in predicting the moduli. A review covering the properties of random fibre composites has been written by Bert (1979). Most estimates are based on an approach introduced by Nielsen & Chen (1968) which assumes that the modulus $\bar{E}$ can be obtained by integrating the expression

$$\bar{E} = 2\pi \int_0^{\pi/2} E(\theta)\, \mathrm{d}\theta \tag{5.24}$$

where $E(\theta)$ is the orientation dependence of the elastic modulus of a unidirectional lamina, with the same V_f such that

$$\frac{1}{E(\theta)} = \frac{1}{E_\parallel} c^4 + \left(\frac{1}{G_\#} - \frac{2\nu_{\parallel\perp}}{E_\parallel}\right) c^2 s^2 + \frac{1}{E_\perp} s^4 \tag{5.25}$$

where $c = \cos\theta$ and $s = \sin\theta$. More details of equation (5.25) can be found in Section 6.2. Akasaka (1974) derived the expressions

$$\bar{E} = \left[\frac{E_\parallel + E_\perp + 2\nu_{\parallel\perp} E_\perp}{1 - \nu_{\parallel\perp}\nu_{\perp\parallel}}\right] \times \left[\frac{E_\parallel + E_\perp - 2\nu_{\parallel\perp} E_\perp + 4(1 - \nu_{\parallel\perp}\nu_{\perp\parallel}) G_\#}{3(E_\parallel + E_\perp) + 2\nu_{\parallel\perp} E_\perp + 4(1 - \nu_{\parallel\perp}\nu_{\perp\parallel}) G_\#}\right] \tag{5.26}$$

$$\bar{G} = \frac{E_\parallel + E_\perp - 2\nu_{\parallel\perp} E_\perp}{8(1 - \nu_{\parallel\perp}\nu_{\perp\parallel})} + \tfrac{1}{2} G_\# \tag{5.27}$$

$$\bar{\nu} = \frac{\bar{E}}{2\bar{G}} - 1 \tag{5.28}$$

Fig. 5.9. Orientation dependence of Young's modulus of glass fibre–polyester resin ($V_f = 0.30$).

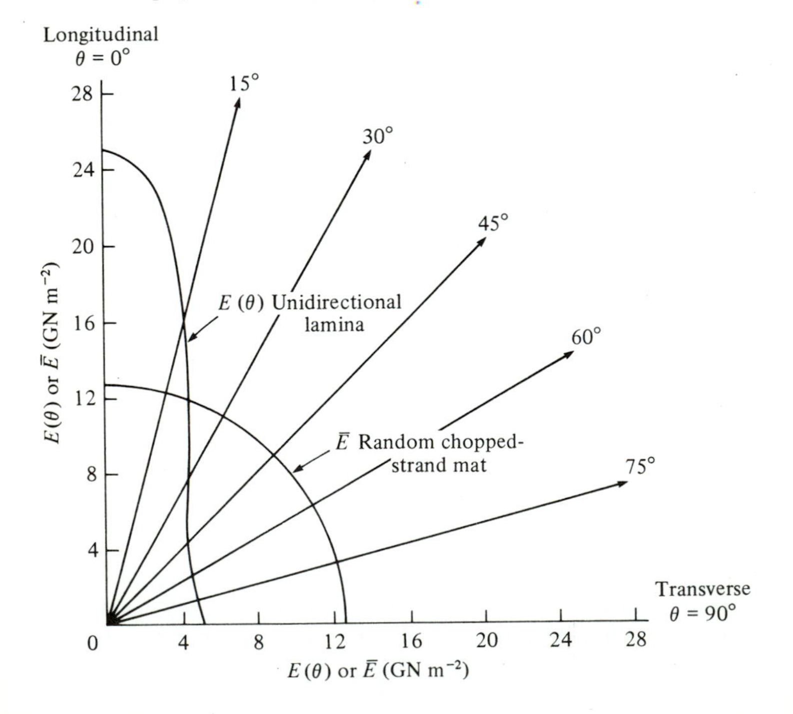

Equations (5.25) and (5.26) are too complicated for most practical cases and the following simplified equations may be used:

$$\bar{E} = \tfrac{3}{8}E_{\parallel} + \tfrac{5}{8}E_{\perp} \tag{5.29}$$

$$\bar{G} = \tfrac{1}{8}E_{\parallel} + \tfrac{1}{4}E_{\perp} \tag{5.30}$$

The effect of volume fraction of fibres in these equations is accounted for in the dependence of $E_{\parallel}$ and $E_{\perp}$ on V_f. The through thickness elastic properties of in-plane random fibre laminae will be considerably less than the in-plane properties because there are no fibres aligned in the through-thickness direction.

Equations (5.26) to (5.30) refer to completely random fibre distributions. It is now possible to tailor-make laminae with pre-determined fibre orientation distributions so that the elastic properties as well as other physical and mechanical properties can be designed to meet specific requirements.

By way of summarising the main ideas in Sections 5.2 and 5.3 the variation of Young's modulus with the direction of the tensile axis for unidirectional laminae and in-plane random laminae fabricated from a glass fibre–polyester resin, $V_f = 0.30$, is shown in Fig. 5.9. The curves for unidirectional laminae were calculated from equations (5.19, 5.21, 5.22, 5.25) and for in-plane random laminae from equations (5.29) using the following experimentally determined parameters. $E_f = 76$ GN m^{-2}, $E_m = 3.5$ GN m^{-2}, $\xi = 0.2$ and 1.0 for $E_{\perp}$ and $G_{\#}$ respectively, and $\nu_{\parallel\perp} = 0.27$. G_f and G_m were determined from equation (6.3) using values of $\nu_f = 0.25$ and $\nu_m = 0.37$ respectively. Some typical values of $E_{\parallel}$, $E_{\perp}$, $G_{\#}$ and $\nu_{\parallel\perp}$ for glass fibre–polyester resin, carbon fibre–epoxy resin and Kevlar 49 fibre–epoxy resin are given in Table 5.1. With these data plots similar to those in Fig. 5.9 can be obtained.

Table 5.1. *Typical values of elastic constants of unidirectional laminae,* $V_f \approx 0.50$

Material	$E_{\parallel T} \approx E_{\parallel C}$ (GN m^{-2})	$E_{\perp}$ (GN m^{-2})	$G_{\#}$ (GN m^{-2})	$\nu_{\parallel\perp}$
Glass–polyester	35–40	8–12	3.5–5.5	0.26
Type I carbon–epoxy	190–240	5–8	3–6	0.26
Kevlar 49–epoxy	65–75	4–5	2–3	0.35

Note: T, tension; C, compression.

5.4 Stress and strain distribution at fibre ends

In the interpretation of elastic properties in Sections 5.2 and 5.3 it is assumed that either the fibres do not end except at the surfaces of the laminae or that the fibres are so long that effects associated with fibre ends can be neglected. However, as the aspect ratio $l/2r$ of the fibres decreases, the effect of fibre ends becomes progressively more significant since the stress and strain fields in the fibre and in the surrounding matrix are modified by the discontinuity. The 'efficiency' of the fibres in stiffening and reinforcing the matrix decreases as the fibre length decreases. Fibre ends play an important role in the fracture of short fibre composites and also in continuous fibre composites since the long fibres may break down into discrete lengths.

Consider a fibre of length l embedded in a matrix of lower modulus as illustrated in Fig. 5.10. If the fibre is well bonded to the matrix an

Fig. 5.10. (*a*) Diagrammatic representation of deformation around a discontinuous fibre embedded in a matrix subjected to a tensile load parallel to the fibre. (*b*) Variation along the fibre of tensile stress in the fibre and shear stress at the interface according to Cox (1952). Matrix and fibre remain elastic and the interface bond is perfect. l_c is the critical value of the fibre length for the maximum stress.

(*a*)

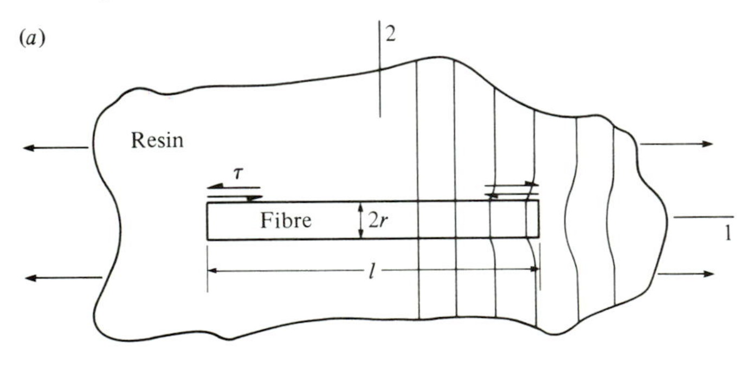

(*b*)

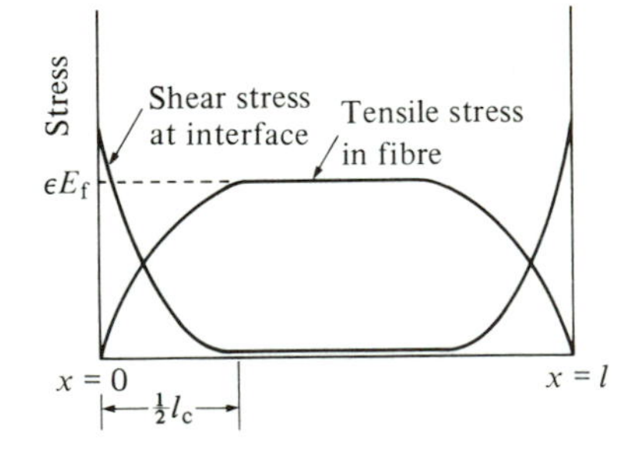

applied stress on the resin will be transferred to the fibre across the interface. In the region of the fibre ends the strain in the fibre will be less than in the matrix. The deformation field in the matrix has been sketched in Fig. 5.10*a*. The problem has been treated analytically by Cox (1952) (see Holister & Thomas (1966) for a full review), using the so-called *shear-lag analysis*, with the assumptions that both the fibre and the matrix remain elastic, and the interface satisfies the assumptions listed in Section 3.1. Cox showed that for an applied stress on the resin parallel to the fibre the tensile stress along the fibre is given by

$$\sigma_f = E_f \epsilon_m \left\{ \frac{1 - \cosh \beta(\frac{1}{2}l - x)}{\cosh \frac{1}{2}\beta l} \right\} \tag{5.31}$$

where $$\beta = \left\{ \frac{2G_m}{E_f r^2 \ln (R/r)} \right\}^{\frac{1}{2}} \tag{5.32}$$

and $2R$ is the interfibre spacing. Thus, the tensile stress is zero at the fibre ends ($x = 0$) and is a maximum in the centre of the fibre, as illustrated schematically in Fig. 5.10*b*. The shear stress at the interface is given by

$$\tau = E_f \epsilon_m \left\{ \frac{G_m}{2E_f \ln (R/r)} \right\}^{\frac{1}{2}} \frac{\sinh \beta(\frac{1}{2}l - x)}{\cosh \frac{1}{2}\beta l} \tag{5.33}$$

The form of this expression is also shown in Fig. 5.10*b*. The shear stress is a maximum at the fibre ends and falls almost to zero in the centre. These results show that there are regions at the ends of the fibre which do not carry the full load so that the average stress in a fibre of length l is less than in a continuous fibre subjected to the same external loading conditions. The reinforcing efficiency decreases as the average fibre length decreases because a greater proportion of the total fibre length is not fully loaded. The maximum possible value of the strain in the fibre is the strain ϵ applied to the composite material as a whole so that the maximum stress in the fibre is ϵE_f. To achieve this maximum stress the fibre length must be greater than a critical value l_c. The schematic representation in Fig. 5.10*b* shows that for fibres longer than l_c the regions at the ends of the fibre which are not fully loaded have a length $\frac{1}{2}l_c$.

It will be evident that the reinforcing efficiency is dependent on the strength of the interface since the load transfer requires a strong interface bond. The large shear stresses at the ends of fibres can result in (i) shear debonding at the interface, (ii) cohesive failure of the matrix (iii) cohesive failure of the fibre and (iv) shear yielding of the matrix, depending on the relative failure strengths associated with these processes. The other factor limiting the reinforcing efficiency is the strength of the fibre. The ratio of the maximum interface shear stress at the fibre ends to the maximum tensile stress in the centre of the fibre can be calculated from equations (5.31)–(5.33) and is

$$\frac{\tau_{\max}}{\sigma_{\mathrm{f\,max}}} = \left\{\frac{G_{\mathrm{m}}}{2E_{\mathrm{f}}\ln(R/r)}\right\}^{\frac{1}{2}} \coth \tfrac{1}{4}\beta l \tag{5.34}$$

which, for long fibres, reduces to

$$\frac{\tau_{\max}}{\sigma_{\mathrm{f\,max}}} = \left\{\frac{G_{\mathrm{m}}}{2E_{\mathrm{f}}\ln(R/r)}\right\}^{\frac{1}{2}} \tag{5.35}$$

Some values of $\tau_{\max}/\sigma_{\mathrm{f\,max}}$ are given in Table 5.2 for four composite materials. Further discussion of these equations is included in Chapter 7 on fracture mechanisms. It should be noted that the analysis used here is not exact. Finite element analysis and experimental studies suggest that it under-estimates the shear stress concentration at the ends of the fibres by at least a factor of two.

5.5 Elastic properties of short fibre composite materials

Since the reinforcing efficiency of short fibres is less than that of long fibres, it follows that the effective modulus of short fibre composite materials will be less also. In the general case a material has a three-dimensional distribution of fibre orientations (Section 4.7) and a distribution of fibre lengths (Section 4.6). There is no satisfactory description of the elastic properties in terms of these parameters and in this section only a few simple ideas will be considered.

For a unidirectionally aligned material containing fibres of length l

Table 5.2. *Values of* $\tau_{\max}/\sigma_{\mathrm{f\,max}}$ *for some typical composite materials*

Material	$G_{\mathrm{m}}/E_{\mathrm{f}}$	V_{f}	$\tau_{\max}/\sigma_{\mathrm{f\,max}}$
Glass–polyester	0.017	0.3	0.13
	0.017	0.6	0.25
Type I carbon–epoxy	0.005	0.3	0.07
	0.005	0.6	0.13
Glass–nylon	0.010	0.3	0.10
	0.010	0.6	0.19
Kevlar 49–epoxy	0.014	0.3	0.12
	0.014	0.6	0.23

Note: A square array of fibres is assumed.

the rule of mixtures equation (5.11) may be modified by the inclusion of a length correction factor, η_l, so that

$$E_{\parallel} = \eta_l E_f V_f + E_m(1 - V_f) \tag{5.36}$$

Cox (1952) derived the following expression for η_l.

$$\eta_l = 1 - (\tanh \tfrac{1}{2}\beta l)/\tfrac{1}{2}\beta l \tag{5.37}$$

where β is given by equation (5.32). Combination of equations (5.11), (5.36) and (5.37) shows that when $E_f \gg E_m$ the effective modulus of the short fibre material compared with continuous fibre material is

$$E(\text{short})/E(\text{continuous}) = \eta_l \tag{5.38}$$

Some values of η_l calculated for fibre lengths 0.1, 1.0 and 10 mm for two composite systems are given in Table 5.3. This shows that providing there is a strong interface bond the efficiency of stiffness reinforcement by short fibres is close to 100% provided the fibre length is over 1.0 mm. An example from work by Dingle (1974) on aligned discontinuous carbon fibre–epoxy composite materials prepared by a novel method is set out in Table 5.4. No significant increase in efficiency is obtained by increasing the fibre length from 1 mm to 6 mm. Imperfect alignment of the short fibres probably explains the reduced efficiency compared with continuous fibres.

An alternative form of equation (5.36), which was developed by Halpin (1969) as an extension of the Halpin–Tsai equations, expresses the longitudinal Young's modulus as

$$E_{\parallel} = \frac{E_m(1 + \xi\eta V_f)}{(1 - \eta V_f)} \tag{5.39}$$

where η is given by equation (5.22) and $\xi = l/r$.

When there is a distribution of fibre orientations the reinforcing

Table 5.3. *Values of length correction factor η_l for carbon–epoxy and glass–nylon composite materials*

Material	l(mm)	G_m/E_f	r(μm)	V_f	η_l
Carbon–epoxy	0.1	0.005	8	0.3	0.20
	1.0	0.005	8	0.3	0.89
	10.0	0.005	8	0.3	0.99
Glass–nylon	0.1	0.010	11	0.3	0.21
	1.0	0.010	11	0.3	0.89
	10.0	0.010	11	0.3	0.99

efficiency of the fibres is further reduced; Cox (1952) included an additional term in equation (5.36) to take this into account, thus

$$E = \eta_o \eta_l E_f V_f + E_m(1 - V_f) \tag{5.40}$$

where η_o is an orientation efficiency factor.

Values of η_o have been calculated by Krenchel (1964) for different fibre orientation distributions assuming that the matrix and fibre deform elastically and the strains are equal. Thus, a group of parallel fibres with a total cross-sectional area Δa_f lying at an angle θ to the applied load are equivalent to a group of fibres of area $\Delta a_f'$ aligned in the direction of the applied load, where

$$\Delta a_f' = \Delta a_f \cos^4 \theta \tag{5.41}$$

For groups of differently oriented fibres the equivalent area of the total reinforcement is

$$a_f' = \Sigma \Delta a_f \cos^4 \theta \tag{5.42}$$

The orientation efficiency factor η_0 is defined by

$$\eta_0 = a_f'/a_f = \Sigma \Delta a_f \cos^4 \theta / a_f \tag{5.43}$$

For unidirectional laminae $\eta_0 = 1$ and 0 when tested parallel and perpendicular to the fibres respectively. $\eta_0 = \frac{3}{8}$ for in-plane random fibre distributions and $\eta_0 = \frac{1}{5}$ for three-dimensional random distributions. This approach is an alternative to the methods described in Section 5.3. This subject has been developed further by a number of workers and is particularly relevant to the prediction of the properties of injection moulded products where a range of fibre orientation distributions and fibre length distributions may exist. Some of the theoretical and experimental details have been reviewed by Darlington & Smith (1978).

Table 5.4. *Young's modulus of aligned discontinuous carbon fibre–epoxy resin composite materials (from Dingle 1974)*

Fibre length, l (mm)	V_f	$E_\parallel$ predicted, continuous fibres (GN m^{-2})	$E_\parallel$ experimental, discontinuous fibres (GN m^{-2})	η_l
1	0.49	194	155	0.80
4	0.32	128	112	0.87
6	0.42	167	141	0.84

Note: $E_\parallel$ was calculated assuming $E_f = 390$ GN m^{-2}.

5.6 Thermal stresses and curing stresses

Shrinkage stresses during cure and thermal stresses due to differences between the thermal expansion coefficients of the matrix and fibre may have a major effect on the microstresses within a composite material and are additional to the stresses produced by external loads. The microstresses are often sufficient to produce microcracking even in the absence of external loads.

The stresses due to curing arise from a combination of resin shrinkage during the curing processes and differential thermal contraction after post-curing at an elevated temperature. Much of the resin shrinkage occurs while it is still liquid so that stresses do not develop. This shrinkage can lead to sink marks and other undesirable surface effects on plastic products. For this reason alone a considerable effort has been applied to developing resin systems which have low shrinkage characteristics. Resin shrinkage in the later stages of cure leads to microstresses which cannot be relieved. The level of these stresses depends sensitively on the microstructural parameters discussed in Chapter 4. Some simple examples will be considered.

For a single fibre surrounded by resin (Fig. 5.11*a*) shrinkage of the resin results in compressive stresses acting on the fibre normal to the fibre–matrix interface. The stresses will be the same as those generated when a rod of radius $r+\delta r$ is introduced into a hole of radius r. The shrinkage is δr. In contrast, shrinkage of a resin pocket between closely

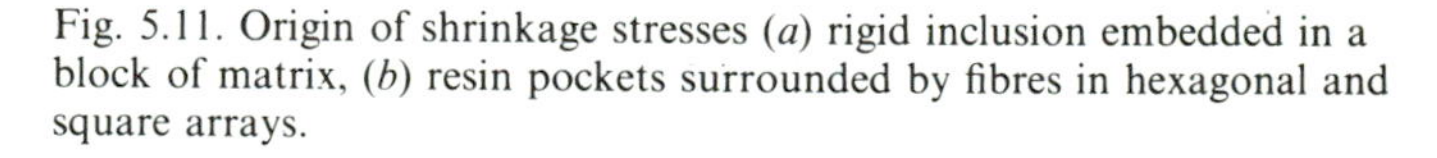

Fig. 5.11. Origin of shrinkage stresses (*a*) rigid inclusion embedded in a block of matrix, (*b*) resin pockets surrounded by fibres in hexagonal and square arrays.

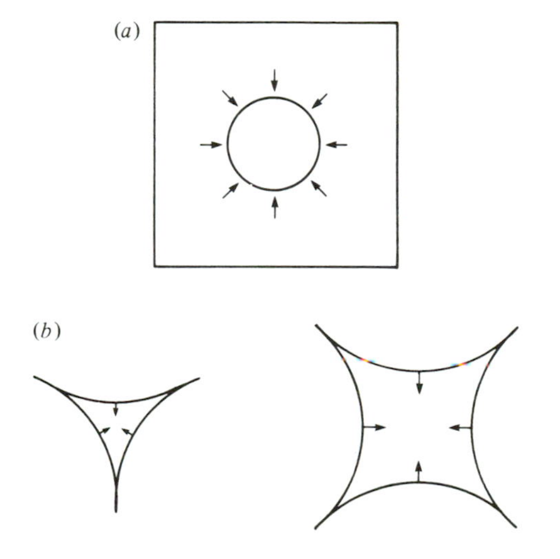

packed fibres (Fig. 5.11*b*) will produce tensile stresses in the resin and across the fibre–matrix interface. The examples in Fig. 5.11*b* correspond to maximum packing density for square and hexagonal arrays of unidirectionally aligned fibres. In real systems, where the fibre distribution is far more random and V_f is less than the maximum, both the effects illustrated in Fig. 5.11 are likely to occur and there will be tensile and compressive microstresses present in the matrix. The occurrence of microstresses is readily demonstrated by photoelastic studies. For example, Koufopoulos & Theocaris (1969) showed that for a square array of circular inclusions in a matrix of lower modulus the microstresses depend on the amount of shrinkage and the elastic moduli of the components. The magnitude of the stresses also depends on V_f, which determines the spacing between the fibres, i.e. $s/2r$, and varies with position within the matrix. Thus, the radial stresses at O (see Fig. 5.12) are tensile and a state of hydrostatic tension exists at this point. Between A and B there is a radial compressive stress which increases with decreasing $s/2r$, or increasing V_f, and a tensile hoop stress. At very small $s/2r$ and high E_f/E_m this tensile stress may become compressive giving a region of hydrostatic compression. The radial stress at C may be tensile or compressive depending on $s/2r$ and E_f/E_m. Tensile stresses are more likely at low V_f and high E_f/E_m and it is possible that these stresses may generate interface cracking. Some of the more detailed analyses of the stresses have been reviewed by Chamis (1974).

For the model shown in Fig. 5.12 and discussed above, there are no external constraints to shrinkage. However, if the lamina is bonded

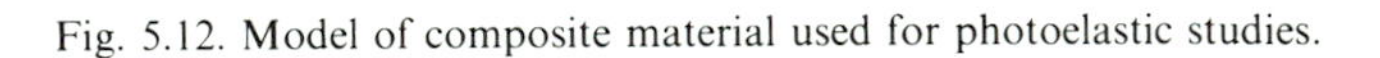

Fig. 5.12. Model of composite material used for photoelastic studies.

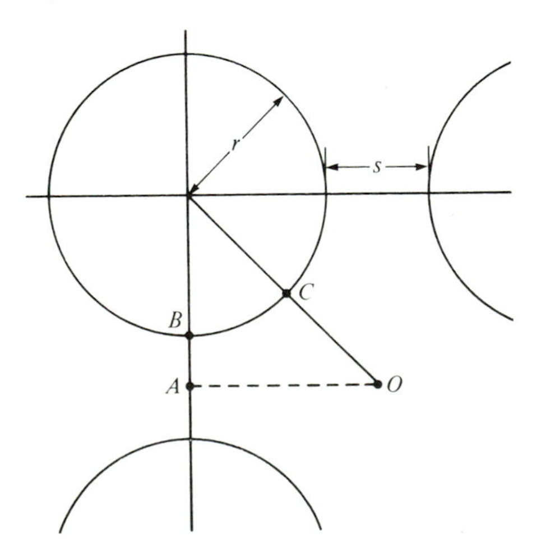

between two other laminae in a laminate, the other laminae constrain the shrinkage deformation and this leads to a different internal stress pattern. In addition shrinkage of the resin results in the fibres being put into compression along their length as illustrated schematically in Fig. 5.13. This increases the tendency for fibre buckling and produces interface shear stresses particularly at fibre ends.

Differential thermal contraction occurs after post-curing at elevated temperatures. In a typical cure cycle polyester-based composite materials are given a final cure at 120 °C followed by slow cooling. For glass fibres and polyester resin with linear coefficients of expansion of 4.7×10^{-6}/°C and 1.5×10^{-4}/°C respectively, cooling to 20 °C produces a differential strain of the order of 1.5% which is about 75% of the strain to failure of resin in uniaxial tension. For a three-dimensional distribution of fibres the stress pattern is very complicated and depends on the distribution and volume fraction of fibres. Since carbon and Kevlar 49 fibres have negative linear coefficients of expansion parallel to their lengths, the thermal microstresses parallel to the fibres are greater than for glass fibres for a given resin system. In contrast the expansion coefficients perpendicular to carbon and Kevlar 49 fibres are positive and this results in corresponding differences in the microstresses.

The strong bonding between fibres and matrix and the different expansion coefficients means that the expansion coefficients of laminae depend on factors such as V_f, fibre arrangement, and E_f/E_m. Expressions have been derived for various fibre geometries. Thus, for example, the linear coefficients of thermal expansion parallel and transverse to the fibres of a unidirectional lamina (Schapery 1968) are given by

$$\alpha_{\parallel} = \frac{E_f \alpha_f V_f + E_m \alpha_m (1 - V_f)}{E_f V_f + E_m (1 - V_f)} \tag{5.44}$$

$$\alpha_{\perp} = (1 + \nu_m) \alpha_m (1 - V_f) + (1 + \nu_f) \alpha_f V_f - \alpha_{\parallel} [\nu_f V_f + \nu_m (1 - V_f)] \tag{5.45}$$

where α_f and α_m are the coefficients of expansion of the fibre and matrix respectively. Laminate theory (Chapter 6) can be used to predict

Fig. 5.13. Shrinkage of matrix, producing compressive stresses along the fibre axis.

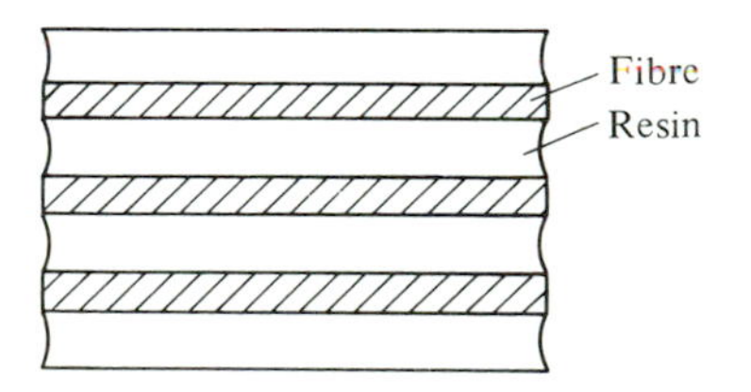

the expansion coefficients of laminates from the properties of the laminae.

Similarly, Halpin (1969) has provided the following expression for the expansion coefficient of a long fibre in-plane random material

$$\bar{\alpha} = \tfrac{1}{2}(\alpha_{\parallel} + \alpha_{\perp}) + \frac{\tfrac{1}{2}(E_{\parallel} - E_{\perp})}{E_{\parallel} + (1 + 2\nu_{\parallel\perp})\,E_{\perp}}(\alpha_{\parallel} - \alpha_{\perp}) \tag{5.46}$$

All the values in this equation relate to the same V_{f}.

References and further reading

Adams, D. F. & Doner, D. R. (1967). Transverse normal loading of a unidirectional composite. *J. Comp. Mater.* **1**, 152–64.

Akasaka, T. (1974) A practical method of evaluating the isotropic elastic constants of glass mat reinforced plastics. *Comp. Mater. Struct.* (Japan) **3**, 21–2.

Bert, C. W. (1979) Composite-material mechanics: prediction of properties of planar-random fiber composites. *Proceedings of the 34th SRI/RP Conference*, paper 20-A. Society of the Plastics Industry, New York.

Chamis, C. C. (1974) Mechanics of load transfer at the interface. *Composite Materials*, vol. 6, ed. E. P. Plueddemann, pp. 31–77. Academic Press, New York.

Cox, H. L. (1952) The elasticity and strength of paper and other fibrous materials. *Br. J. appl. Phys.* **3**, 72–9.

Darlington, M. W. & Smith, G. R. (1977) Mechanical properties of glass fibre and mineral reinforced polyamides 6 and 6.6. *Reinforced Thermoplastics II*, Plastics and Rubber Institute Conference, paper 12.

Dingle, L. E. (1974) Aligned discontinuous carbon fibre composites. *Proceedings of the Fourth International Conference on Carbon Fibres, their Composites and Applications*, paper 11. Plastics Institute, London.

Holister, G. S. & Thomas, C. (1966) *Fibre Reinforced Materials*. Elsevier, London.

Halpin, J. C. & Tsai, S. W. (1967) Environmental factors in composite materials design. Air Force Materials Laboratory Technical Report AFML-TR-67-423.

Halpin, J. C. (1969) Stiffness and expansion estimates for oriented short fiber composites. *J. Comp. Mater.* **3**, 732–4.

Kelly, A. (1973) *Strong Solids*. Clarendon Press, Oxford.

Kies, J. A. (1962) Maximum strains in the resin of fiberglass composites. US Naval Research Laboratory Report NRL 5752.

Krenchel, H. (1964) *Fibre Reinforcement*. Akademisk Forlag, Copenhagen.

Koufopoulos, T. & Theocaris, P. S. (1969) Shrinkage stresses in two-phase materials. *J. Comp. Mater.* **3**, 308–20.

Lees, J. K. (1968) A study of the tensile modulus of short fiber reinforced plastics. *Polym. Engng Sci.* **8**, 186–94.

Marloff, R. H. & Daniel, I. M. (1969) Three dimensional photoanalysis of a fibre reinforced composite model. *Expl. Mech.* **9**, 156–62.

Nielsen, L. E. & Chen, P. E. (1968) Young's modulus of composites filled with randomly oriented fibres. *J. Mater.* **3**, 352–8.

Puck, A. (1967) Zur Beanspruchung und Verformung von GFK-Mehrschichtenverbund-Bauelementen. *Kunstoffe* **57**, 965–73.
Schapery, J. (1968) Thermal expansion coefficients of composite materials based on energy principles. *J. Comp. Mater.* **2**, 380–404.

6 Laminate theory

6.1 Introduction

In chapter 1 it was shown that almost all engineering materials have a composite character in that on a microscopic scale they consist of a fine dispersion of one phase in another. In considering the elastic properties of bulk materials the microheterogeneity is ignored. Thus, for example, it is usual to treat metals as isotropic and homogeneous. This means that the properties at a point in the body are the same in every direction (isotropic) and they are uniform throughout, being independent of position in the body (homogeneous). Similarly, in the analysis of the elastic properties of laminates, it is assumed that the laminae are homogeneous although microscopic studies reveal considerable evidence for microheterogeneity. In Chapter 5 it was shown that the properties of laminae are not isotropic. The most important class of non-isotropic bodies associated with simple laminate constructions is *orthotropic* bodies which have three mutually perpendicular planes of material symmetry and the properties at any point are different in three mutually perpendicular directions. A unidirectional lamina has three mutually perpendicular planes of symmetry as

Fig. 6.1. Three mutually perpendicular planes of material symmetry in a unidirectional lamina.

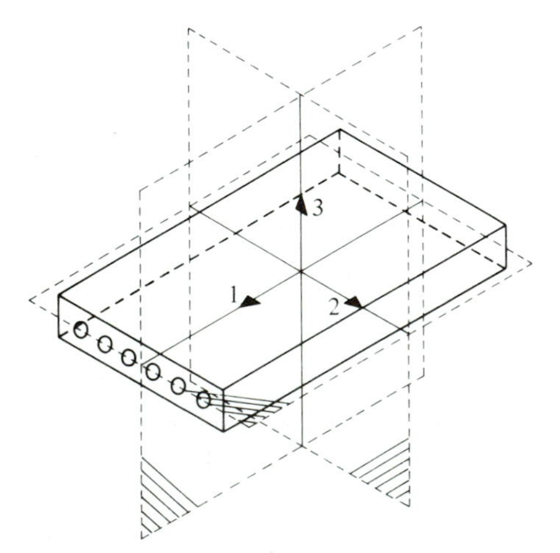

illustrated in Fig. 6.1 and is, therefore, orthotropic. A woven roving lamina is orthotropic and a chopped strand mat lamina is usually assumed to be isotropic in the plane of the lamina.

This chapter describes the methods which are used to calculate the elastic properties of laminates from the properties, orientation and distribution of individual laminae. This is important in the design of composite laminates since it is essential that the response of the final laminate to external loads can be predicted. The approach is based on laminate theory. Full descriptions can be found in many text-books. One of the most straightforward accounts is given by Jones (1975) which makes extensive use of an earlier book by Ashton, Halpin & Petit (1969). The theory uses anisotropic elasticity theory which requires a knowledge of matrix algebra. This chapter begins with a brief introduction to elasticity theory and develops the main steps required to calculate laminate properties. Rigorous proofs and mathematical analyses are avoided and matrix notation is described in sufficient detail that the text can be followed by those not familiar with this subject. Some simple examples are given to demonstrate the way the theory is used. Also in this chapter the methods used to predict the stresses in the individual laminae when a load is applied to a laminate are described. These methods are required for the prediction of laminate failure processes described in Chapter 8.

6.2 Elastic properties of a unidirectional lamina

The first part of this section is included to give some insight into the physical interpretation of the behaviour of non-isotropic bodies. As indicated above this must not be treated as a rigorous analysis and reference should be made to texts on elasticity and laminate theory if the subject is to be developed.

The stresses at a point in a solid can be represented by the stresses acting on the surfaces of a cube at that point using the notation shown in Fig. 6.2. There are three normal stresses σ_{11}, σ_{22} and σ_{33} and three shear stresses τ_{23}, τ_{31} and τ_{12} (the first suffix refers to the direction normal to the plane in which the stress is acting and the second suffix to the direction in which the stress is acting). The corresponding strains are given the notation ϵ_{11}, ϵ_{22}, ϵ_{33}, γ_{23}, γ_{31} and γ_{12}. A number of different notations are used in the literature and care is needed in changing from one notation to another. Thus, for example, a contracted notation replaces σ_{11}, σ_{22}, σ_{33}, τ_{23}, τ_{31} and τ_{12} by σ_1, σ_2, σ_3, σ_4, σ_5 and σ_6 and similarly with ϵ_{11} etc. The stress components shown on the faces of the cube are taken as positive and can be taken to be the forces (per unit area) exerted by the material outside the cube upon the material inside.

Thus, normal tensile stresses are positive, and normal compressive stresses are negative. Nine stress components must be used to define the state of stress at a point, namely σ_1, σ_2, σ_3, τ_{23}, τ_{31}, τ_{12}, τ_{32}, τ_{13}, and τ_{21}. By considering moments of forces about the co-ordinate axes through the centre of the unit cube it can be shown that for equilibrium at any point $\tau_{23} = \tau_{32}$, $\tau_{31} = \tau_{13}$ and $\tau_{12} = \tau_{21}$.

When a unidirectional tensile stress is applied to a solid the elastic strain ϵ in the direction of the applied stress is related to applied stress by the well-known equation

$$\sigma = E\epsilon \tag{6.1}$$

where E is Young's modulus. This is, of course, a simple statement of Hooke's law. The normal strain transverse to the applied stress is $-\nu\epsilon$ where ν is Poisson's ratio. For an isotropic material E and ν are independent of direction of the applied stress. The shear modulus G is defined by

$$\tau = G\gamma \tag{6.2}$$

where γ is the engineering shear strain. For an isotropic material

$$G = E/2(1+\nu) \tag{6.3}$$

Hooke's law can be expressed in a generalised form, using the contracted notation referred to above, as

$$\sigma_i = \sum_{j=1}^{6} C_{ij}\epsilon_j \tag{6.4}$$

Fig. 6.2. Components of stress acting on an elemental unit cube.

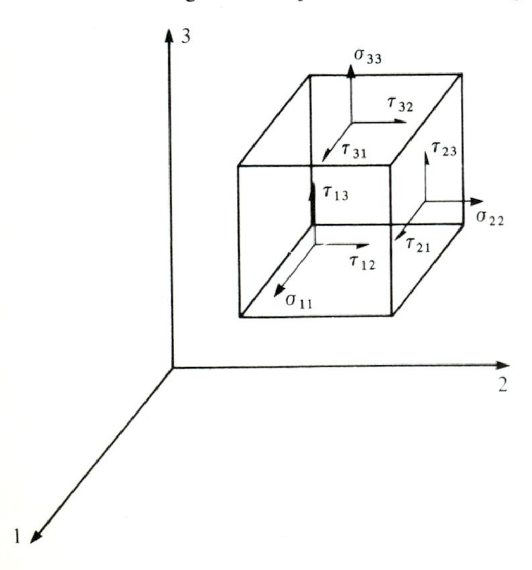

where $i, j = 1, \ldots, 6$. The σ_i are the stress components and the ϵ_j are the strain components. C_{ij} is called the *stiffness* matrix. It can be shown that $C_{ij} = C_{ji}$ so that in an expanded form equation (6.4) becomes

$$\begin{bmatrix} \sigma_1 \\ \sigma_2 \\ \sigma_3 \\ \tau_{23} \\ \tau_{31} \\ \tau_{12} \end{bmatrix} = \begin{bmatrix} C_{11} & C_{12} & C_{13} & C_{14} & C_{15} & C_{16} \\ C_{12} & C_{22} & C_{23} & C_{24} & C_{25} & C_{26} \\ C_{13} & C_{23} & C_{33} & C_{34} & C_{35} & C_{36} \\ C_{14} & C_{24} & C_{34} & C_{44} & C_{45} & C_{46} \\ C_{15} & C_{25} & C_{35} & C_{45} & C_{55} & C_{56} \\ C_{16} & C_{26} & C_{36} & C_{46} & C_{56} & C_{66} \end{bmatrix} \begin{bmatrix} \epsilon_1 \\ \epsilon_2 \\ \epsilon_3 \\ \gamma_{23} \\ \gamma_{31} \\ \gamma_{12} \end{bmatrix} \quad (6.5a)$$

which is the matrix notation for six equations relating stress to strain, the first two being,

$$\left.\begin{aligned} \sigma_1 &= C_{11}\epsilon_1 + C_{12}\epsilon_2 + C_{13}\epsilon_3 + C_{14}\gamma_{23} + C_{15}\gamma_{31} + C_{16}\gamma_{12} \\ \sigma_2 &= C_{12}\epsilon_1 + C_{22}\epsilon_2 + C_{23}\epsilon_3 + C_{24}\gamma_{23} + C_{25}\gamma_{31} + C_{26}\gamma_{12} \end{aligned}\right\} \quad (6.5b)$$

For isotropic materials the stiffness matrix is much simpler because the elastic properties are the same in all directions. Equation (6.5) reduces to

$$\begin{bmatrix} \sigma_1 \\ \sigma_2 \\ \sigma_3 \\ \tau_{23} \\ \tau_{31} \\ \tau_{12} \end{bmatrix} = \begin{bmatrix} C_{11} & C_{12} & C_{12} & 0 & 0 & 0 \\ C_{12} & C_{11} & C_{12} & 0 & 0 & 0 \\ C_{12} & C_{12} & C_{11} & 0 & 0 & 0 \\ 0 & 0 & 0 & \frac{1}{2}(C_{11}-C_{12}) & 0 & 0 \\ 0 & 0 & 0 & 0 & \frac{1}{2}(C_{11}-C_{12}) & 0 \\ 0 & 0 & 0 & 0 & 0 & \frac{1}{2}(C_{11}-C_{12}) \end{bmatrix} \begin{bmatrix} \epsilon_1 \\ \epsilon_2 \\ \epsilon_3 \\ \gamma_{23} \\ \gamma_{31} \\ \gamma_{12} \end{bmatrix} \quad (6.6)$$

There is a corresponding set of equations which relate strain to stress

$$\epsilon_i = \sum_{j=1}^{6} S_{ij}\sigma_j \quad (6.7)$$

where S_{ij} is the *compliance* matrix. For an isotropic material this equation reduces to

$$\begin{bmatrix} \epsilon_1 \\ \epsilon_2 \\ \epsilon_3 \\ \gamma_{23} \\ \gamma_{31} \\ \gamma_{12} \end{bmatrix} = \begin{bmatrix} S_{11} & S_{12} & S_{12} & 0 & 0 & 0 \\ S_{12} & S_{11} & S_{12} & 0 & 0 & 0 \\ S_{12} & S_{12} & S_{11} & 0 & 0 & 0 \\ 0 & 0 & 0 & 2(S_{11}-S_{12}) & 0 & 0 \\ 0 & 0 & 0 & 0 & 2(S_{11}-S_{12}) & 0 \\ 0 & 0 & 0 & 0 & 0 & 2(S_{11}-S_{12}) \end{bmatrix} \begin{bmatrix} \sigma_1 \\ \sigma_2 \\ \sigma_3 \\ \tau_{23} \\ \tau_{31} \\ \tau_{12} \end{bmatrix} \quad (6.8)$$

The compliances can be expressed in terms of the engineering constants and equation (6.8) becomes

$$\begin{bmatrix}\epsilon_1\\ \epsilon_2\\ \epsilon_3\\ \gamma_{23}\\ \gamma_{31}\\ \gamma_{12}\end{bmatrix} = \begin{bmatrix}1/E & -\nu/E & -\nu/E & 0 & 0 & 0\\ -\nu/E & 1/E & -\nu/E & 0 & 0 & 0\\ -\nu/E & -\nu/E & 1/E & 0 & 0 & 0\\ 0 & 0 & 0 & 1/G & 0 & 0\\ 0 & 0 & 0 & 0 & 1/G & 0\\ 0 & 0 & 0 & 0 & 0 & 1/G\end{bmatrix} \begin{bmatrix}\sigma_1\\ \sigma_2\\ \sigma_3\\ \tau_{23}\\ \tau_{31}\\ \tau_{12}\end{bmatrix} \tag{6.9}$$

These equations show that all the relations between stress and strain for isotropic materials can be described in terms of only two independent elastic constants, namely the compliances S_{11} and S_{12}, or the engineering constants E and ν since G is defined by equation (6.3). For the simple uniaxial tensile test referred to at the beginning of this section $\sigma_1 = \sigma$ and $\sigma_2 = \sigma_3 = \tau_{23} = \tau_{31} = \tau_{12} = 0$. Therefore equation (6.9) reduces to

$$\epsilon_1 = (1/E)\,\sigma \tag{6.10}$$

which is the same as equation (6.1), and

$$\epsilon_2 = \epsilon_3 = -(\nu/E)\,\sigma \tag{6.11}$$

In the case of laminae and laminates it is assumed that they are sufficiently thin that the through-thickness stresses are zero. Thus $\sigma_3 = \tau_{23} = \tau_{31} = 0$ (plane stress) and equation (6.9) becomes

$$\begin{bmatrix}\epsilon_1\\ \epsilon_2\\ \gamma_{12}\end{bmatrix} = \begin{bmatrix}1/E & -\nu/E & 0\\ -\nu/E & 1/E & 0\\ 0 & 0 & 1/G\end{bmatrix} \begin{bmatrix}\sigma_1\\ \sigma_2\\ \tau_{12}\end{bmatrix} \tag{6.12}$$

for isotropic materials. Similarly, if we replace the stiffness coefficients by engineering constants, equation (6.6) becomes

$$\begin{bmatrix}\sigma_1\\ \sigma_2\\ \tau_{12}\end{bmatrix} = \begin{bmatrix}E/(1-\nu^2) & \nu E/(1-\nu^2) & 0\\ \nu E/(1-\nu^2) & E/(1-\nu^2) & 0\\ 0 & 0 & G\end{bmatrix} \begin{bmatrix}\epsilon_1\\ \epsilon_2\\ \gamma_{12}\end{bmatrix} \tag{6.13}$$

The orthotropic lamina shown in Fig. 6.1 can be assumed to be isotropic in the plane normal to the 1-axis since the properties are independent of direction in that plane. The strain–stress equation of the lamina assuming plane stress conditions is

$$\begin{bmatrix}\epsilon_1\\ \epsilon_2\\ \gamma_{12}\end{bmatrix} = \begin{bmatrix}S_{11} & S_{12} & 0\\ S_{12} & S_{22} & 0\\ 0 & 0 & S_{66}\end{bmatrix} \begin{bmatrix}\sigma_1\\ \sigma_2\\ \tau_{12}\end{bmatrix} \tag{6.14}$$

and $\quad S_{11} = 1/E_1, \quad S_{22} = 1/E_2, \quad S_{66} = 1/G_{12}$ (6.15)

and $\quad S_{12} = -\nu_{12}/E_1 = -\nu_{21}/E_2$ (6.16)

The *reciprocal relation* equation (6.16) requires some explanation. Poisson's ratio ν_{12} refers to the strains produced in the 2-direction when a test sample is stressed in the 1-direction and is given by $-\epsilon_2/\epsilon_1$. Similarly $\nu_{21} = -\epsilon_1/\epsilon_2$.

The stress–strain relation for a unidirectional lamina is

$$\begin{bmatrix} \sigma_1 \\ \sigma_2 \\ \tau_{12} \end{bmatrix} = \begin{bmatrix} Q_{11} & Q_{12} & 0 \\ Q_{12} & Q_{22} & 0 \\ 0 & 0 & Q_{66} \end{bmatrix} \begin{bmatrix} \epsilon_1 \\ \epsilon_2 \\ \gamma_{12} \end{bmatrix} \tag{6.17}$$

Q_{11}, Q_{12} and Q_{66} are called reduced stiffnesses, i.e.

$$\left.\begin{aligned} Q_{11} = C_{11} = \frac{E_1}{1-\nu_{12}\nu_{21}}, \quad Q_{12} = C_{12} = \frac{\nu_{12}E_2}{1-\nu_{12}\nu_{21}} = \frac{\nu_{21}E_1}{1-\nu_{12}\nu_{21}}, \\ Q_{22} = C_{22} = \frac{E_2}{1-\nu_{12}\nu_{21}}, \quad Q_{66} = \tfrac{1}{2}(C_{11}-C_{12}) = G_{12} \end{aligned}\right\} \tag{6.18}$$

There are only four independent constants in these equations i.e. the reduced stiffnesses Q_{11}, Q_{12}, Q_{22} and Q_{66} or the engineering constants E_1, E_2, ν_{12} and ν_{21}. Equation (6.17) indicates that orthotropic materials tested in tension or compression along the principal material directions exhibit no shear strains with respect to these principle directions and the deformation is independent of G_{12}. Similarly, a shear stress τ_{12} produces only shear strains which are independent of E_2, ν_{12} and ν_{21}. In other words there is no *coupling* between tensile and shear strains. This does not apply when the lamina is tested at arbitrary angles to the principal material directions. Thus, consider a lamina tested in such a way that the new co-ordinate system x–y is at an angle θ to the principal material directions as illustrated in Fig. 6.3. Elasticity theory shows that the stress–strain relation becomes

$$\begin{bmatrix} \sigma_x \\ \sigma_y \\ \tau_{xy} \end{bmatrix} = \begin{bmatrix} \bar{Q}_{11} & \bar{Q}_{12} & \bar{Q}_{16} \\ \bar{Q}_{12} & \bar{Q}_{22} & \bar{Q}_{26} \\ \bar{Q}_{16} & \bar{Q}_{26} & \bar{Q}_{66} \end{bmatrix} \begin{bmatrix} \epsilon_x \\ \epsilon_y \\ \gamma_{xy} \end{bmatrix} \tag{6.19}$$

The matrix $\bar{Q}_{ij}$ is called the transformed reduced stiffness matrix and the stiffnesses have the following values

Fig. 6.3. Rotation of axes from co-ordinate system 1–2 to x–y.

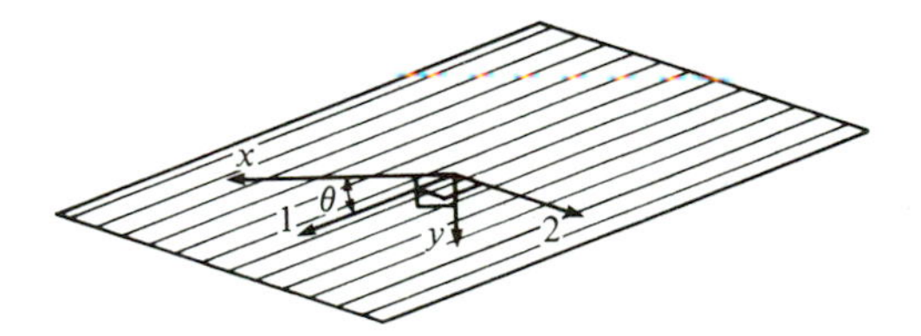

$$\left.\begin{aligned}
\bar{Q}_{11} &= Q_{11}c^4 + 2(Q_{12}+2Q_{66})s^2c^2 + Q_{22}s^4 \\
\bar{Q}_{12} &= (Q_{11}+Q_{22}-4Q_{66})s^2c^2 + Q_{12}(s^4+c^4) \\
\bar{Q}_{22} &= Q_{11}s^4 + 2(Q_{12}+2Q_{66})s^2c^2 + Q_{22}c^4 \\
\bar{Q}_{26} &= (Q_{11}-Q_{12}-2Q_{66})sc^3 + (Q_{12}-Q_{22}+2Q_{66})s^3c \\
\bar{Q}_{26} &= (Q_{11}-Q_{12}-2Q_{66})s^3c + (Q_{12}-Q_{22}+2Q_{66})sc^3 \\
\bar{Q}_{66} &= (Q_{11}+Q_{22}-2Q_{12}-2Q_{66})s^2c^2 + Q_{66}(s^4+c^4)
\end{aligned}\right\} \qquad (6.20)$$

where s and c are $\sin\theta$ and $\cos\theta$ respectively. There is a corresponding equation for the strain–stress relation

$$\begin{bmatrix} \epsilon_x \\ \epsilon_y \\ \gamma_{xy} \end{bmatrix} = \begin{bmatrix} \bar{S}_{11} & \bar{S}_{12} & \bar{S}_{16} \\ \bar{S}_{12} & \bar{S}_{22} & \bar{S}_{26} \\ \bar{S}_{16} & \bar{S}_{26} & \bar{S}_{66} \end{bmatrix} \begin{bmatrix} \sigma_x \\ \sigma_y \\ \tau_{xy} \end{bmatrix} \qquad (6.21)$$

and $\bar{S}_{ij}$ can be represented by a set of equations similar to equations (6.20). Since $\bar{S}_{16}$ and $\bar{S}_{26}$ are not zero, it follows that a unidirectional stress $\sigma_x(\sigma_y = \tau_{xy} = 0)$ will produce both normal and shear strains, whereas as observed previously a unidirectional stress σ_1 does not produce a shear strain with respect to the principle directions.

Using the approach outlined above, it is possible to obtain expressions for the elastic properties E_x, E_y, G_{xy} and ν_{xy} corresponding to the x–y co-ordinate system. These are

$$\left.\begin{aligned}
\frac{1}{E_x} &= \frac{1}{E_1}c^4 + \left(\frac{1}{G_{12}} - \frac{2\nu_{12}}{E_1}\right)s^2c^2 + \frac{1}{E_2}s^4 \\
\frac{1}{E_y} &= \frac{1}{E_1}s^4 + \left(\frac{1}{G_{12}} - \frac{2\nu_{12}}{E_1}\right)s^2c^2 + \frac{1}{E_2}c^4 \\
\frac{1}{G_{xy}} &= 2\left(\frac{2}{E_1} + \frac{2}{E_2} + \frac{4\nu_{12}}{E_1} - \frac{1}{G_{12}}\right)s^2c^2 + \frac{1}{G_{12}}(s^4+c^4) \\
\nu_{xy} &= E_x\left[\frac{\nu_{12}}{E_1}(s^4+c^4) - \left(\frac{1}{E_1} + \frac{1}{E_2} - \frac{1}{G_{12}}\right)s^2c^2\right]
\end{aligned}\right\} \qquad (6.22)$$

Thus, if E_1, E_2, G_{12} and ν_{12} are known the elastic properties at any angle can be calculated. As an example, consider a unidirectional lamina of glass fibre and polyester resin with $V_f = 0.5$. The properties of this lamina related to the 1–2 co-ordinate system are given in Table 5.1: $E_1 = E_\parallel = 40$ GN m^{-2}, $E_2 = E_\perp = 8.2$ GN m^{-2}, $G_{12} = G_\# = 3.9$ GN m^{-2}, and $\nu_{12} = \nu_{\parallel\perp} = 0.26$.† It is a straightforward matter using a programmable pocket calculator to substitute these values in equations (6.22) and then plot out the variation of $E_x/E_\parallel$, $E_y/E_\parallel$, $G_{xy}/G_\#$ and ν_{xy} as a function of θ as shown in Fig. 6.4. The general form of the curves in Fig. 6.4 is independent of $E_\perp/E_\parallel$, but the maximum and minimum

† This notation is used in the remainder of the book and is applicable to unidirectional lamina only.

Fig. 6.4. Moduli of a glass fibre–polyester lamina ($V_f = 0.5$) (*a*) $E_x/E_\parallel$ and $E_y/E_\parallel$ and (*b*) $G_{xy}/G_\#$ and ν_{xy}.

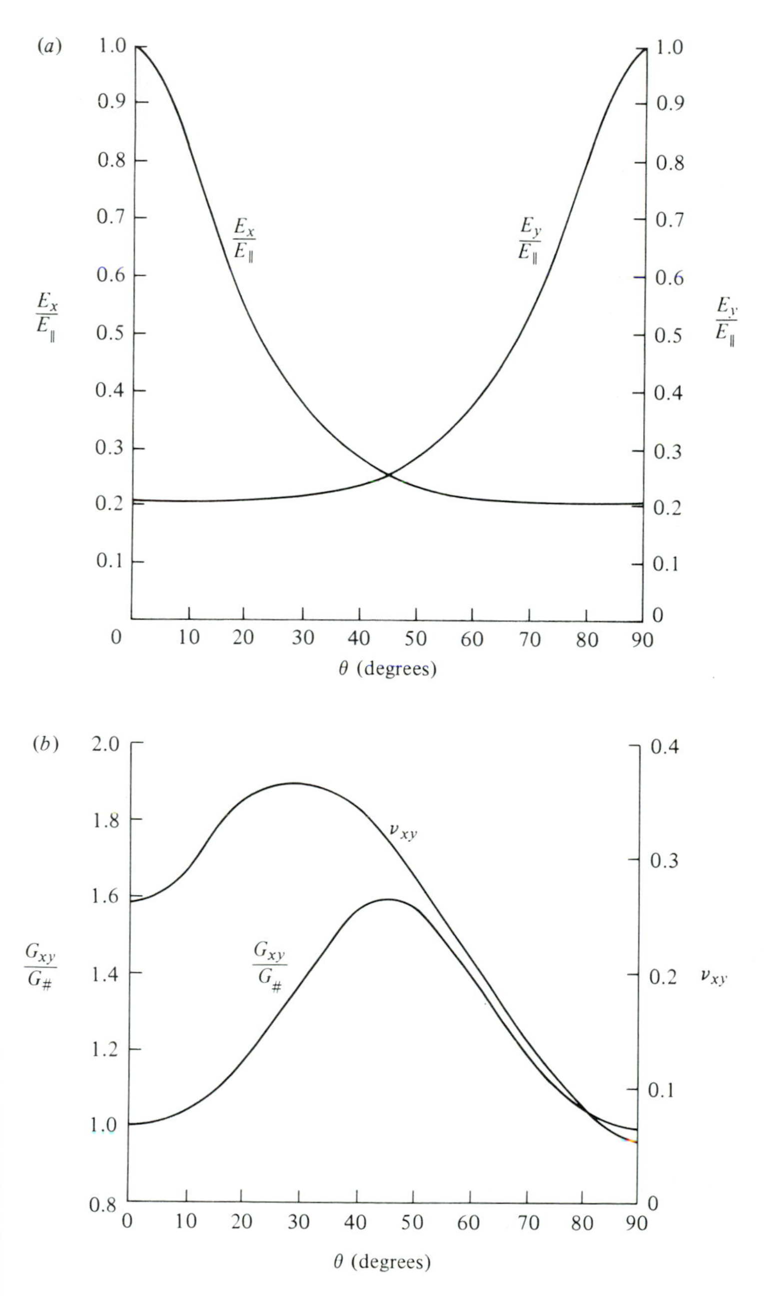

values of E_x do not necessarily occur at $\theta = 0°$ and $\theta = 90°$ respectively.

Equation (6.22) can be tested experimentally using off-axis tensile tests as described by Sinclair & Chamis (1979). Test specimens are cut at a range of angles from a flat sheet of unidirectionally aligned fibres as illustrated in Fig. 6.5. The strains are measured using strain gauges bonded to the centre of the test sample. Great care is necessary to minimise stress concentrations at the ends of the specimens and obtain reproducible results. The results obtained for a Type I carbon fibre in an epoxy resin matrix ($V_f \approx 0.5$) are shown in Fig. 6.6. These are compared with the predictions of equation (6.22) using experimentally determined values of $E_{\parallel}$, $E_{\perp}$, $G_{\#}$ and $\nu_{\parallel\perp}$ (i.e. $E_{\parallel} = 241$ GN m^{-2}, $E_{\perp} = 7.7$ GN m^{-2}, $G_{\#} = 6.1$ GN m^{-2}, and $\nu_{\parallel\perp} = 0.27$). There is good agreement between the predicted and experimental values. Similar agreement was obtained for other elastic constants.

6.3 Elastic properties of laminates

Fig. 6.6 highlights a serious limitation of unidirectional fibre composites, which is the very large in-plane anisotropy. To overcome this problem it is necessary to lay-up the unidirectional laminae at

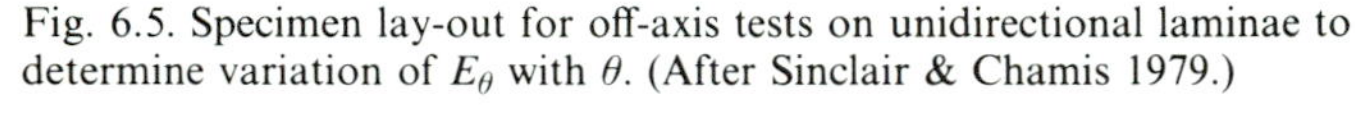

Fig. 6.5. Specimen lay-out for off-axis tests on unidirectional laminae to determine variation of E_θ with θ. (After Sinclair & Chamis 1979.)

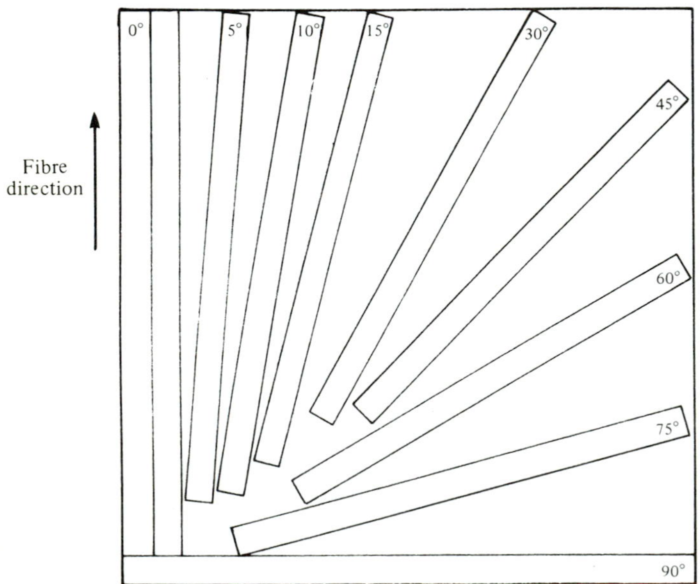

different angles as outlined in Section 4.1. The elastic properties of the laminate depend on the properties of the laminae. In this section an outline is given of the approach which is used to predict laminate properties based on classical lamination theory. As in the previous section rigorous mathematical detail is avoided.

A number of assumptions are made in lamination theory to obtain theoretical predictions. These are (i) the laminae are perfectly bonded and do not slip relative to each other, (ii) the bond between the laminae is infinitely thin, and (iii) the laminate has the properties of a thin sheet. With these assumptions it is possible to treat the laminate as a thin elastic plate and apply the classical analysis of Kirchhoff to derive the strain distribution throughout the plate when it is subjected to external forces. Since the laminate is made up of laminae oriented in different directions with respect to each other, but having the same stress–strain relations given by equation (6.19) (assuming that the laminate is not a hybrid; see Section 10.5) the stress–strain equation of the kth layer of the laminate can be written

$$\begin{bmatrix} \sigma_x \\ \sigma_y \\ \tau_{xy} \end{bmatrix}_k = \begin{bmatrix} \bar{Q}_{11} & \bar{Q}_{12} & \bar{Q}_{16} \\ \bar{Q}_{12} & \bar{Q}_{22} & \bar{Q}_{26} \\ \bar{Q}_{16} & \bar{Q}_{26} & \bar{Q}_{66} \end{bmatrix}_k \begin{bmatrix} \epsilon_x \\ \epsilon_y \\ \gamma_{xy} \end{bmatrix}_k \tag{6.23}$$

Fig. 6.6. Moduli E_θ of a carbon fibre–epoxy resin unidirectional lamina ($V_f \approx 0.50$) tested at various angles to the fibre direction. (After Sinclair & Chamis 1979.)

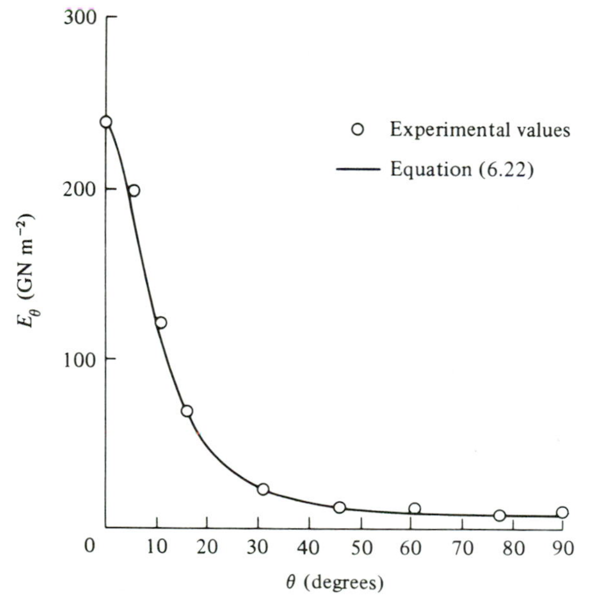

Clearly, the $\overline{Q}_{ij}$ have to be evaluated for each layer since it is dependent on the angle between the co-ordinate system used and the principal material directions, i.e. θ in Fig. 6.3. Thus for a given strain distribution the stress in each layer can be evaluated. Classical lamination theory provides a method of calculating the resultant forces and moments per unit length acting on the laminate by integrating the stresses acting in each lamina through the thickness of the laminate. Specific expressions for the stiffnesses of the laminate can be derived in terms of the laminate construction i.e. the orientation and thickness of the individual laminae, the stacking sequence and the elastic constants of each laminae, $E_{\parallel}$, $E_{\perp}$, $G_{\#}$ and $\nu_{\parallel\perp}$. By way of example, the stiffnesses of two common and relatively simple laminate constructions are given later in this section.

Before considering these examples it is important to have a physical

Fig. 6.7. Two layer cross-ply laminate (*a*) geometry (*b*) illustration of different strains produced when a fixed stress is applied to layers individually and to bonded laminate.

(*a*)

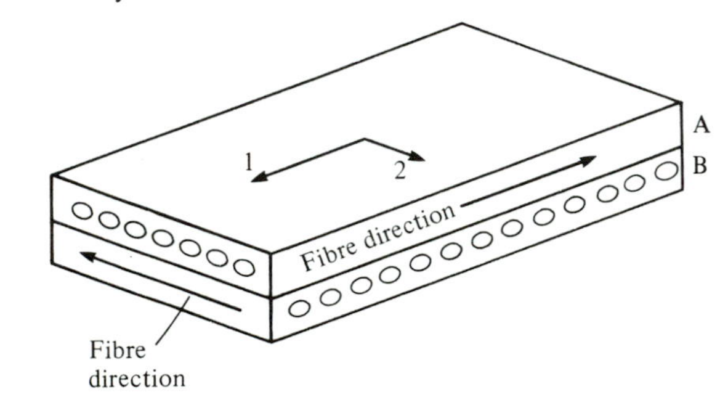

(*b*)

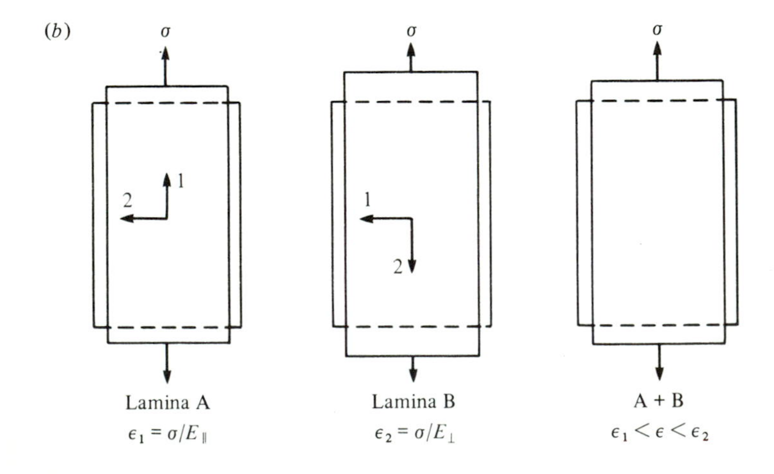

insight into the nature of the coupling stresses which arise when two or more non-isotropic, in this case orthotropic, laminae are bonded together and then stretched. Consider the cross-ply laminate in Fig. 6.7 which consists of two unidirectional laminae bonded together with the fibre directions at right angles to each other. Because $E_{\parallel} \neq E_{\perp}$ the application of a unidirectional tensile stress σ, as shown to each lamina separately produces different tensile strains $\sigma/E_{\parallel}$ and $\sigma/E_{\perp}$ parallel to the stress direction. When the laminae are bonded together as in a thin plate, the applied stress produces the *same* strain in each lamina providing it remains flat and the laminate has a new tensile modulus E_{L}, where $E_{\parallel} > E_{\mathrm{L}} > E_{\perp}$. Clearly, the tensile stresses in each lamina are not the same and this leads to coupling forces which produce additional stresses in the plane normal to the tensile axis. The existence of such forces can be demonstrated experimentally in two ways. Firstly, take a well-bonded two layer cross-ply laminate and heat it. Since the linear coefficients of thermal expansion $\alpha_{\parallel}$ and $\alpha_{\perp}$ are not the same the laminate will behave as a bimetallic strip (Section 1.1) and bend towards the lamina with the lower coefficient of expansion, see Fig. 6.8. Secondly, take two laminae and apply a fixed *strain* to each, parallel and perpendicular to the fibre direction respectively. Because $E_{\parallel} \neq E_{\perp}$ different stresses will be required to produce the same strain in each lamina. While the laminae are still under stress, bond the two together and then release the stress. Again the laminate will bend and additional stresses normal to the plane of the laminate are required to keep it flat.

The same arguments apply to a two layer angle-ply laminate as illustrated in Fig. 6.9. The application of a tensile stress to each lamina separately at an angle to the fibre direction produces a change of shape which involves both extensional and shear strains as will be evident from equation (6.21). The shape change in each layer is the same but of opposite sense. When the two laminae are bonded together the shape change is quite different, as shown, and for a two layer laminate out-of-plane coupling stresses develop when the laminate is stressed.

Fig. 6.8. Double curvature distortion produced by heating a two layer cross-ply laminate.

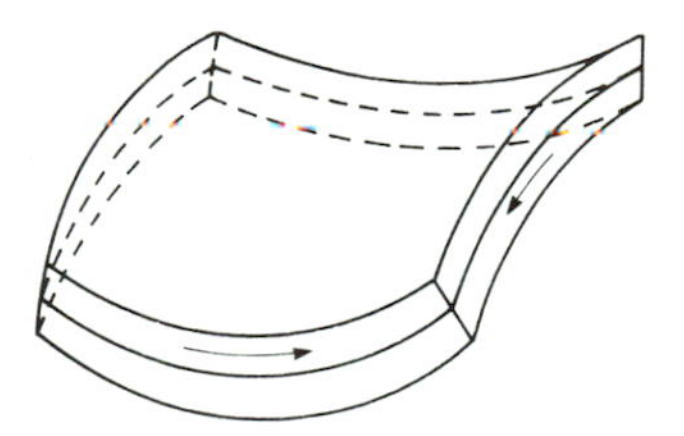

The same experiments described above can be repeated although the second experiment is difficult to perform because of the complicated shape change. Heating of the well-bonded laminate results in a twisting as illustrated in Fig. 6.10. Similarly, if the well-bonded laminate is subjected to a fixed load without any end constraints to keep the laminate flat as in Fig. 6.9, the coupling forces lead to a twist also.

The coupling stresses arise because the laminate is not symmetrical

Fig. 6.9. Shape change in single lamina at $+\theta$ and $-\theta$ and a two lamina $\pm\theta$ angle-ply laminate subjected to unidirectional tensile stress.

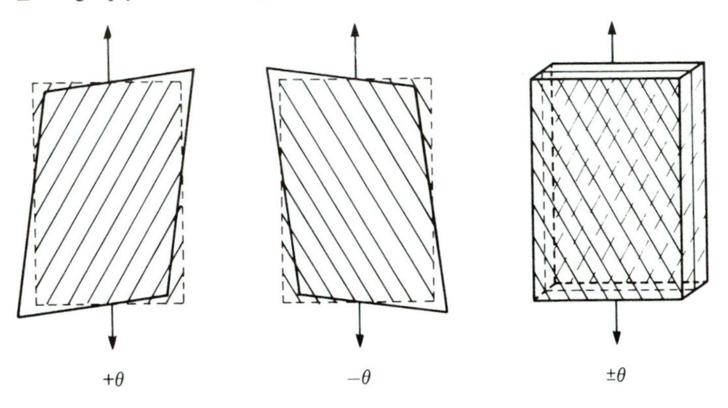

Fig. 6.10. Shape change of a non-symmetric angle-ply laminate on heating.

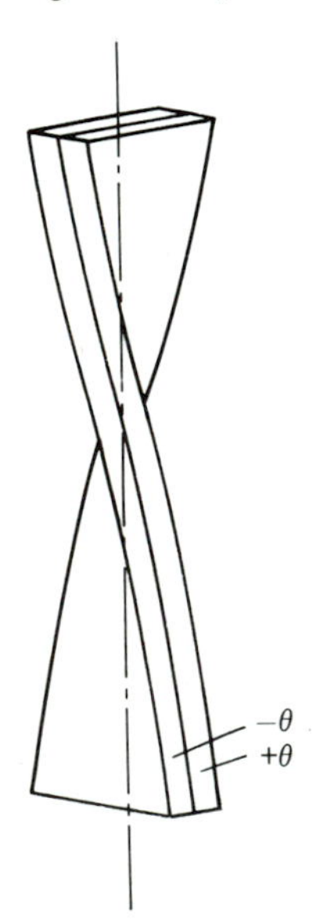

about the centre plane and they add an additional degree of complexity to the prediction of the response of a laminate to applied external forces. For this, and other reasons, it is common practice in many applications to use symmetric laminates which are not subject to this type of coupling. The symmetry depends on the number and thickness of the differently oriented laminae about the centre plane. Some examples of symmetric and non-symmetric laminates are shown in Fig. 6.11. For a simple non-symmetric $\pm\theta$ laminate with uniform thickness laminae, as illustrated in Fig. 6.11*b*, the coupling stresses decrease as the number of layers increases.

Symmetric cross-ply laminates

Consider the laminate structure illustrated in Fig. 6.11*c* consisting of three layers thickness t_1 oriented in one direction and two layers thickness t_2 oriented in the other direction. The ratio of the total

Fig. 6.11. Examples of symmetric and non-symmetric angle-ply and cross-ply laminates.

Symmetric laminates

(*a*) $+\theta$ t; $-\theta$ t; $-\theta$ t; $+\theta$ t

$+\theta$ t_1; $-\theta$ t_2; $+\theta$ t_1; $+\theta$ t_1; $-\theta$ t_2; $+\theta$ t_1

Non-symmetric laminates

(*b*) $+\theta$ t; $-\theta$ t; $+\theta$ t; $-\theta$ t; $+\theta$ t; $-\theta$ t

$+\theta$ t; $-\theta$ t; $-\theta$ t; $+\theta$ $2t$

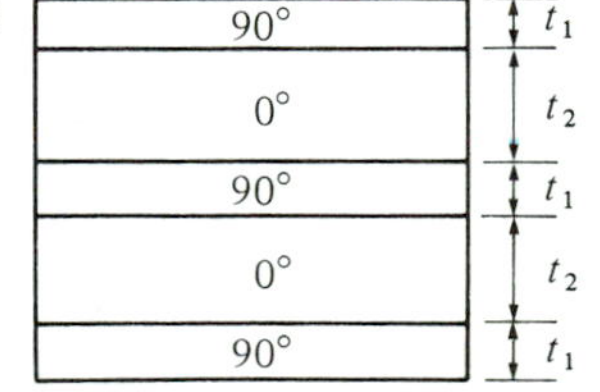

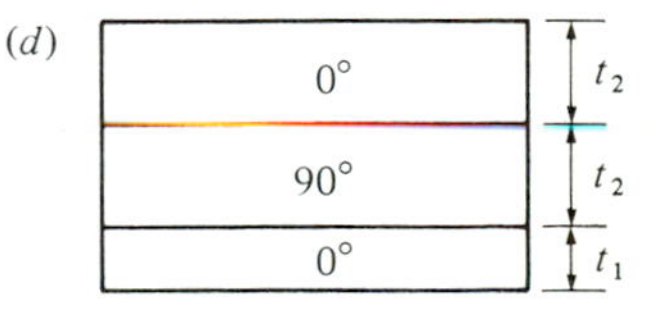

Cross-ply

thicknesses of each set of layers is $3t_1/2t_2$. In the general case this ratio can be expressed as the volume fraction ratio of the two sets of layers V_A/V_B.

The cross-ply laminate has orthotropic elastic properties since there are three mutually perpendicular planes of material symmetry as shown in Fig. 6.12. Thus, the stress–strain relations will have the form of equation (6.17) when the laminate is tested in the principal directions normal to the planes of symmetry, i.e.

$$\begin{bmatrix} \sigma_{c\,1} \\ \sigma_{c\,2} \\ \tau_{c\,12} \end{bmatrix} = \begin{bmatrix} Q_{c\,11} & Q_{c\,12} & Q_{c\,16} \\ Q_{c\,12} & Q_{c\,22} & Q_{c\,26} \\ Q_{c\,16} & Q_{c\,26} & Q_{c\,66} \end{bmatrix} \begin{bmatrix} E_{c\,1} \\ E_{c\,2} \\ \gamma_{c\,12} \end{bmatrix} \tag{6.24}$$

where $Q_{c\,16} = Q_{c\,26} = 0$ for the orthotropic case (note that subscript c refers to cross-ply laminates). Following equation (6.23) the reduced stiffness matrix $Q_{c\,ij}$ in equation (6.24) is obtained by summing the stiffness of the individual layers, i.e.

$$Q_{c\,ij} = \sum_{k=1}^{N} [\bar{Q}_{ij}]_k V_k \tag{6.25}$$

Bearing in mind the volume fraction ratios referred to above and the

Fig. 6.12. Three mutually perpendicular planes of material symmetry in a cross-ply laminate.

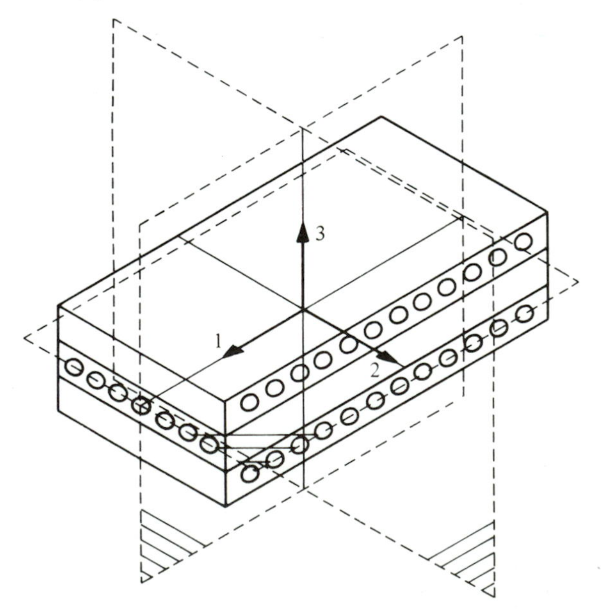

geometry of the laminate, it follows that the values of $Q_{c\,11}$, $Q_{c\,12}$, $Q_{c\,22}$ and $Q_{c\,66}$ are

$$\left.\begin{aligned} Q_{c\,11} &= V_A Q_{11} + V_B Q_{22} \\ Q_{c\,12} &= V_A Q_{12} + V_B Q_{12} = Q_{12} \\ Q_{c\,22} &= V_A Q_{22} + V_B Q_{11} \\ Q_{c\,66} &= V_A Q_{66} + V_B Q_{66} = Q_{66} \end{aligned}\right\} \tag{6.26}$$

where Q_{11}, Q_{12}, Q_{22} and Q_{66} are given by equation (6.18). These equations are, of course, similar to the rule of mixtures equations in Chapter 5 which are used to predict the moduli of laminae in terms of the moduli of the fibres and matrix.

Balanced symmetric angle-ply laminates

Exactly the same arguments as above can be used to obtain the reduced stiffness matrix $Q_{a\,ij}$ (subscript a for angle-ply laminates) since a symmetric angle-ply laminate also has orthotropic elastic properties. Fig. 6.13 shows that the fibre orientations in the laminae are arranged at $\pm\theta$ to the principal direction 1. The stiffnesses are

$$\left.\begin{aligned} Q_{a\,11} &= \bar{Q}_{11}, \quad Q_{a\,12} = \bar{Q}_{12} \\ Q_{a\,22} &= \bar{Q}_{22}, \quad Q_{a\,66} = \bar{Q}_{66} \end{aligned}\right\} \tag{6.27}$$

where $\bar{Q}_{11}$, $\bar{Q}_{12}$, $\bar{Q}_{22}$ and $\bar{Q}_{66}$ are given by equation (6.20). This takes account of the dependence of $Q_{a\,ij}$ on the value of θ. It will be noted that there are no terms $Q_{a\,16}$ and $Q_{a\,26}$ in equation (6.27). This applies only to the special case of balanced and symmetric angle-ply laminates. By referring to equation (6.20), it is seen that the sign of $\bar{Q}_{16}$ and $\bar{Q}_{26}$

Fig. 6.13. Symmetry axes in an angle-ply laminate.

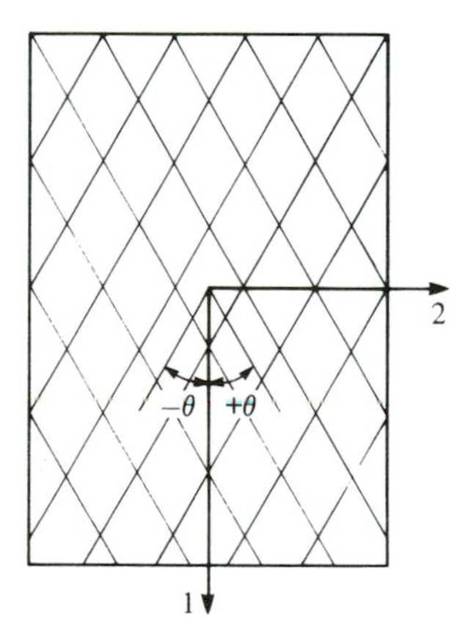

depends on the angle θ. In a balanced laminate the contributions of each layer to Q_{a16} and Q_{a26} cancel out because the total thickness of all the $+\theta$ layers is the same as the total thickness of all the $-\theta$ layers. In Fig. 6.11*a* the four layer laminate is balanced and symmetric and the six layer laminate is balanced when $t_2 = 2t_1$. The volume fraction of the

Fig. 6.14. Elastic constants of $\pm\theta$ angle-ply laminates. (*a*) Type I high modulus carbon fibre–epoxy resin. (*b*) Kevlar 49 fibre–epoxy resin. (Data from Chamis 1980.)

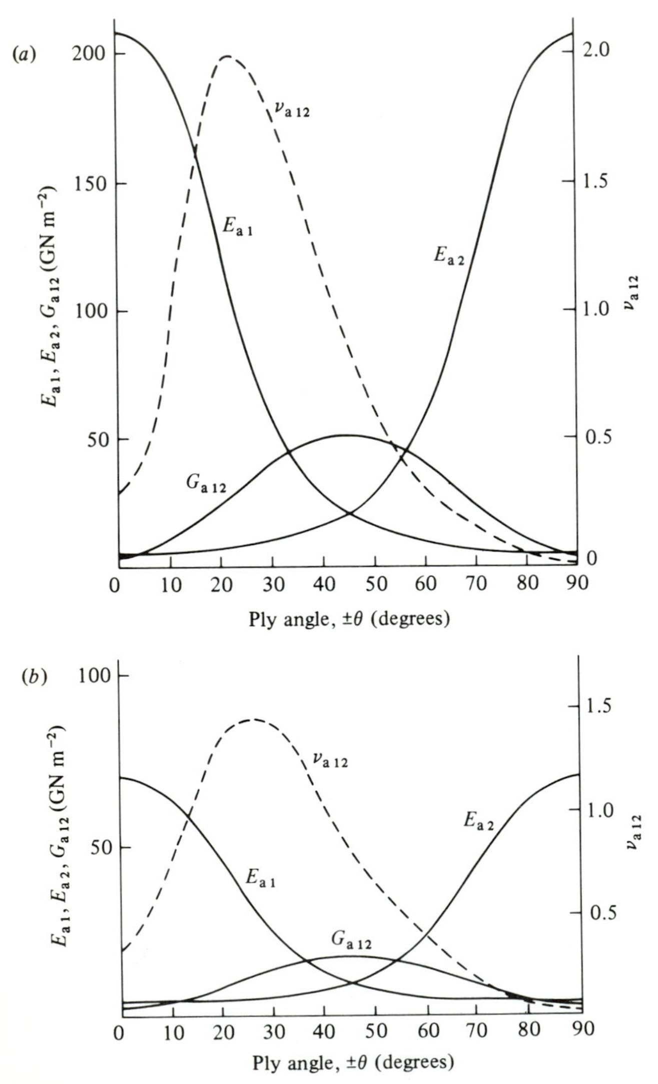

θ layers does not affect Q_{a11}, Q_{a22}, Q_{a12} and Q_{a66} because each layer contributes the same stiffness to angle-ply laminates tested in the principal material directions.

The stiffness of cross-ply and angle-ply laminates in equations (6.26) and (6.27) respectively can also be expressed in terms of the engineering elastic properties of the laminate E_{a1}, E_{a2}, ν_{a12} and G_{a12} and have the same form as equation (6.18). For angle-ply laminates,

$$\left.\begin{aligned} Q_{a11} &= \frac{E_{a1}}{1-\nu_{a12}\nu_{a21}} \\ Q_{a12} &= \frac{\nu_{a12}E_{a2}}{1-\nu_{a12}\nu_{a21}} = \frac{\nu_{a21}E_{a1}}{1-\nu_{a12}\nu_{a21}} \\ Q_{a22} &= \frac{E_{a2}}{1-\nu_{a12}\nu_{a21}} \\ Q_{a66} &= G_{a12} \end{aligned}\right\} \tag{6.28}$$

Thus, the reduced stiffness elastic properties can be calculated from the variation of E_{a1}, E_{a2}, ν_{a12} and G_{a12} with angle θ. An example of the variation of these constants with θ for Kevlar 49 fibre–epoxy resin and for high modulus carbon fibre–epoxy resin is given in Fig. 6.14. Obviously at $\theta = 0°$, $E_{a1} = E_{\parallel}$, $E_{a2} = E_{\perp}$, $\nu_{a12} = \nu_{\parallel\perp}$ and $G_{a12} = G_{\#}$ and these correspond to the values given in Table 5.1. The maximum value of G_{a12} occurs at $\theta = \pm 45°$.

6.4 Determination of stresses acting on individual laminae

To predict the probable failure mechanisms in a laminate (as will be discussed in Chapter 8) it is necessary to be able to determine the stresses, acting in the laminae, which are likely to cause failure. Three main failure processes have been identified, viz. (i) fibre fracture, (ii) transverse tensile cracking parallel to the fibres, and (iii) shear fracture parallel to the fibres. It is reasonable to assume, following the maximum stress criterion for failure (Section 7.8), that in a single lamina there are critical values of the stress for these failure modes. Thus $\sigma_{\parallel}^*$ is the critical stress parallel to the fibres for fibre fracture, and $\sigma_{\perp}^*$ and $\tau_{\#}^*$ are the corresponding stresses for transverse and shear cracking respectively. The values of $\sigma_{\parallel}$, $\sigma_{\perp}$ and $\tau_{\#}$ in individual laminae within a laminate can be calculated from laminate theory as follows. The stress–strain equation of the laminate (equation (6.24)) is inverted to the strain–stress equation

$$\begin{bmatrix} \epsilon_{c1} \\ \epsilon_{c1} \\ \gamma_{c12} \end{bmatrix} = \begin{bmatrix} S_{c11} & S_{c12} & S_{c16} \\ S_{c12} & S_{c22} & S_{c26} \\ S_{c16} & S_{c26} & S_{c66} \end{bmatrix} \begin{bmatrix} \sigma_{c1} \\ \sigma_{c2} \\ \tau_{c12} \end{bmatrix} \tag{6.29}$$

For cross-ply and angle-ply laminates, with laminae of constant thickness, tested with the principal stresses parallel to the directions of symmetry as illustrated in Figs. 6.12 and 6.13. $\tau_{c\,12} = 0$ and $S_{c\,16} = S_{c\,26} = 0$, therefore,

$$\left.\begin{aligned} \epsilon_{c\,1} &= S_{c\,11}\sigma_{c\,1} + S_{c\,12}\sigma_{c\,2} \\ \epsilon_{c\,2} &= S_{c\,12}\sigma_{c\,1} + S_{c\,22}\sigma_{c\,2} \\ \gamma_{c\,12} &= 0 \end{aligned}\right\} \tag{6.30}$$

The strains $\epsilon_{c\,1}$, $\epsilon_{c\,2}$ and $\gamma_{c\,12}$ have to be expressed in terms of the strains parallel and transverse to the fibre i.e. $\epsilon_{\|}$, $\epsilon_{\perp}$ and $\gamma_{\#}$ by using the transformation equations. These can be derived by considering a sheet of material subjected to plane stress ($\sigma_z = 0$). Stresses in the x–y co-ordinate system are transformed to stresses in the x'–y' co-ordinate system by resolving the forces acting on the planes normal to different axes. For example, a uniaxial tensile stress σ_x acting along the x-direction produces a tensile stress $\sigma_x \cos^2\theta$ in the x'-direction, a tensile stress $\sigma_x \sin^2\theta$ in the y'-direction and a shear stress $\sigma_x \sin\theta\cos\theta$ on the plane normal to the y'-direction. The full set of equations for the transformation of stress is

$$\begin{bmatrix} \sigma_x \\ \sigma_y \\ \tau_{xy} \end{bmatrix} = \begin{bmatrix} \cos^2\theta & \sin^2\theta & -2\sin\theta\cos\theta \\ \sin^2\theta & \cos^2\theta & +2\sin\theta\cos\theta \\ \sin\theta\cos\theta & -\sin\theta\cos\theta & \cos^2\theta - \sin^2\theta \end{bmatrix} \begin{bmatrix} \sigma_{x'} \\ \sigma_{y'} \\ \tau_{xy'} \end{bmatrix} \tag{6.31}$$

Similarly, the strain transformation equations are

$$\begin{bmatrix} \epsilon_x \\ \epsilon_y \\ \tfrac{1}{2}\gamma_{xy} \end{bmatrix} = \begin{bmatrix} \cos^2\theta & \sin^2\theta & -2\sin\theta\cos\theta \\ \sin^2\theta & \cos^2\theta & +2\sin\theta\cos\theta \\ \sin\theta\cos\theta & -\sin\theta\cos\theta & \cos^2\theta - \sin^2\theta \end{bmatrix} \begin{bmatrix} \epsilon_{x'} \\ \epsilon_{y'} \\ \tfrac{1}{2}\gamma_{xy'} \end{bmatrix} \tag{6.32}$$

or

$$\begin{bmatrix} \epsilon_{x'} \\ \epsilon_{y'} \\ \tfrac{1}{2}\gamma_{xy'} \end{bmatrix} = \begin{bmatrix} \cos^2\theta & \sin^2\theta & 2\sin\theta\cos\theta \\ \sin^2\theta & \cos^2\theta & -2\sin\theta\cos\theta \\ -\sin\theta\cos\theta & \sin\theta\cos\theta & \cos^2\theta - \sin^2\theta \end{bmatrix} \begin{bmatrix} \epsilon_x \\ \epsilon_y \\ \tfrac{1}{2}\gamma_{xy} \end{bmatrix} \tag{6.33}$$

Thus, for cross-ply laminates, $\theta = 0°$ and $90°$ in alternate laminae so that in laminae in which the 1-direction is parallel to the fibres

$${}^1\epsilon_{\|} = \epsilon_{c\,1}, \quad {}^1\epsilon_{\perp} = \epsilon_{c\,2}, \quad \text{and} \quad {}^1\gamma_{\#} = \gamma_{c\,12} \tag{6.34}$$

and for laminae in which the 2-direction is parallel to the fibres

$${}^2\epsilon_{\|} = \epsilon_{c\,2}, \quad {}^2\epsilon_{\perp} = \epsilon_{c\,1} \quad \text{and} \quad {}^2\gamma_{\#} = \gamma_{c\,12} \tag{6.35}$$

The stresses in the individual laminae can be obtained from the stress–strain equations (6.17) and (6.18) of the laminae, thus for the laminae with the 1-direction parallel to the fibres (equation (6.34))

$$\left.\begin{aligned} {}^1\sigma_{\parallel} &= Q_{11}\epsilon_{c1}+Q_{12}\epsilon_{c2} \\ {}^1\sigma_{\perp} &= Q_{12}\epsilon_{c1}+Q_{22}\epsilon_{c2} \\ {}^1\tau_{\#} &= Q_{66}\gamma_{c12} \end{aligned}\right\} \qquad (6.36)$$

A similar set of equations is obtained from the laminae with the fibres parallel to the 2-direction. Combining equations (6.30) and (6.36) gives

$$\left.\begin{aligned} {}^1\sigma_{\parallel} &= Q_{11}(S_{c11}\sigma_{c1}+S_{c12}\sigma_{c2})+Q_{12}(S_{c12}\sigma_{c1}+S_{c22}\sigma_{c2}) \\ {}^1\sigma_{\perp} &= Q_{12}(S_{c11}\sigma_{c1}+S_{c12}\sigma_{c2})+ Q_{22}(S_{c12}\sigma_{c1}+S_{c22}\sigma_{c2}) \\ {}^1\tau_{\#} &= 0 \end{aligned}\right\} \qquad (6.37)$$

For unidirectional testing $\sigma_{c2} = 0$, therefore

$$\left.\begin{aligned} {}^1\sigma_{\parallel} &= (Q_{11}S_{c11}+Q_{12}S_{c12})\sigma_{c1} \\ {}^1\sigma_{\perp} &= (Q_{12}S_{c11}+Q_{22}S_{c12})\sigma_{c1} \end{aligned}\right\} \qquad (6.38)$$

S_{c11}, S_{c12}, S_{c22} and S_{c66} can be obtained from equation (6.24) by inversion and

$$\left.\begin{aligned} S_{c11} &= \frac{Q_{c22}}{Q_{c11}Q_{c22}-Q_{c12}^2} \\ S_{c12} &= \frac{-Q_{c12}}{Q_{c11}Q_{c22}-Q_{c12}^2} \\ S_{c22} &= \frac{Q_{c11}}{Q_{c11}Q_{c22}-Q_{c12}^2} \end{aligned}\right\} \qquad (6.39)$$

Combining equations (6.18), (6.26), (6.38) and (6.39) and taking $V_A = V_B = \frac{1}{2}$ gives

$$\left.\begin{aligned} \frac{{}^1\sigma_{\parallel}}{\sigma_{c1}} &= \frac{\frac{1}{2}E_1(E_1+E_2)-\nu_{12}^2E_2^2}{\frac{1}{4}(E_1+E_2)^2-\nu_{12}^2E_2^2} \\ \frac{{}^1\sigma_{\perp}}{\sigma_{c1}} &= \frac{\frac{1}{2}\nu_{12}E_2(E_1-E_2)}{\frac{1}{4}(E_1+E_2)^2-\nu_{12}^2E_2^2} \end{aligned}\right\} \qquad (6.40)$$

The corresponding equations for ${}^2\sigma_{\parallel}$ and ${}^2\sigma_{\perp}$ are

$$\left.\begin{aligned} \frac{{}^2\sigma_{\parallel}}{\sigma_{c1}} &= \frac{\frac{1}{2}\nu_{12}E_2(E_2-E_1)}{\frac{1}{4}(E_1+E_2)^2-\nu_{12}^2E_2^2} \\ \frac{{}^2\sigma_{\perp}}{\sigma_{c1}} &= \frac{\frac{1}{2}E_2(E_1+E_2)-\nu_{12}^2E_2^2}{\frac{1}{4}(E_1+E_2)^2-\nu_{12}^2E_2^2} \end{aligned}\right\} \qquad (6.41)$$

These equations can now be applied to a cross-ply laminate. Taking $E_1/E_2 = 10$ and $\nu_{12} = 0.25$ gives

$$\frac{{}^1\sigma_{\parallel}}{\sigma_{c1}} = 1.820, \quad \frac{{}^1\sigma_{\perp}}{\sigma_{c1}} = 0.0373, \quad \frac{{}^2\sigma_{\parallel}}{\sigma_{c1}} = -0.0373, \quad \frac{{}^2\sigma_{\perp}}{\sigma_{c1}} = 0.180$$

The physical significance of these results is illustrated in Fig. 6.15 which

Fig. 6.15. Lamina stresses in a symmetric cross-ply laminate due to a unidirectional tensile stress σ_{c1} parallel to the fibre direction in lamina 1.

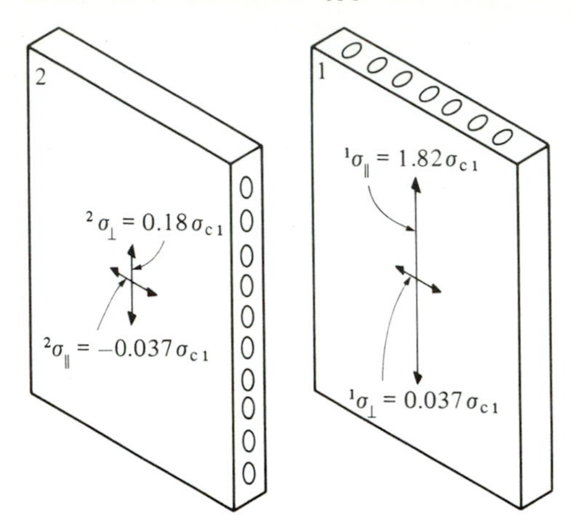

Fig. 6.16. Lamina stresses in a symmetric angle-ply laminate due to a unidirectional tensile stress σ_{a1}. Glass fibre–polyester resin ($V_f = 0.50$). (From Spencer & Hull 1978.)

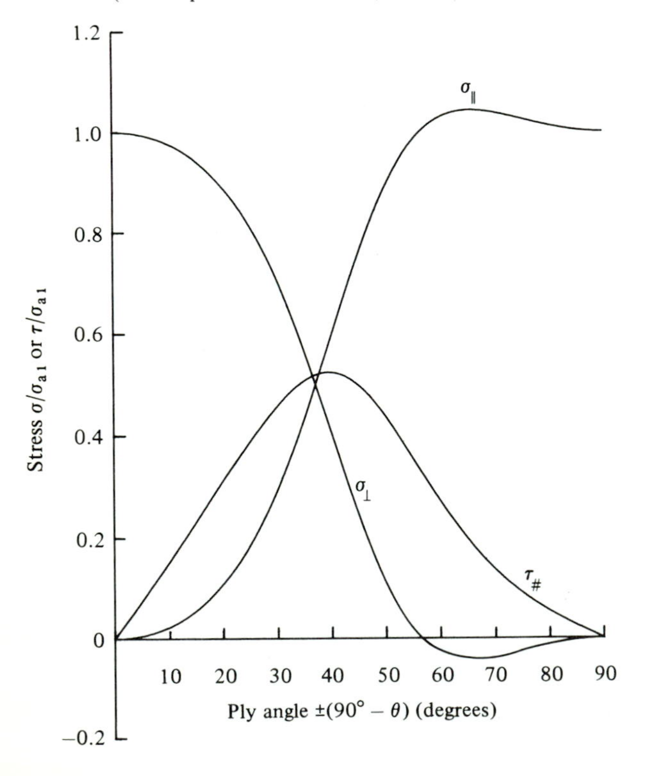

shows two layers of the cross-ply laminate. In lamina 1 the applied stress on the composite, i.e. σ_{c1} produces a stress $1.82\,\sigma_{c1}$ parallel to the fibres. A tensile stress $0.037\,\sigma_{c1}$ is also produced perpendicular to the fibres because of the constraints of the adjacent lamina 2 which prevent the normal Poisson contraction occurring. In lamina 2 the applied stress produces a stress $0.18\,\sigma_{c1}$ perpendicular to the fibres; note that the main part of the load is taken by the fibres in lamina 1 and the Poisson effects lead to a compressive stress parallel to the fibres.

A similar approach can be used to calculate the stresses in angle-ply laminates. The equations are more complicated but a good example can be found in papers by Puck which are summarised by Greenwood (1977). A graphical representation of the stresses calculated for a glass fibre–polyester resin angle-ply laminate ($V_f = 0.50$), tested in uniaxial tension (Fig. 6.13) is shown in Fig. 6.16. The stresses $\sigma_{\parallel}$, $\sigma_{\perp}$ and $\tau_{\#}$ are shown as ratios of σ_{a1}. There is a strong dependence of the stresses on

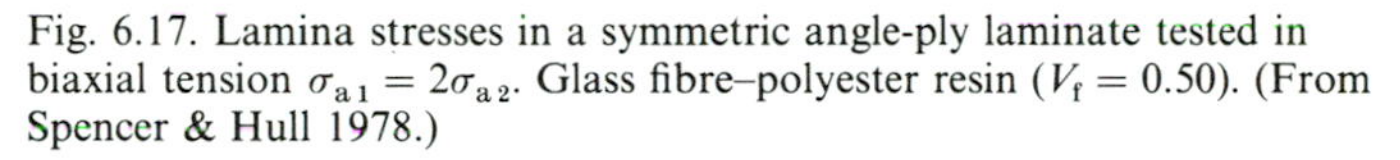

Fig. 6.17. Lamina stresses in a symmetric angle-ply laminate tested in biaxial tension $\sigma_{a1} = 2\sigma_{a2}$. Glass fibre–polyester resin ($V_f = 0.50$). (From Spencer & Hull 1978.)

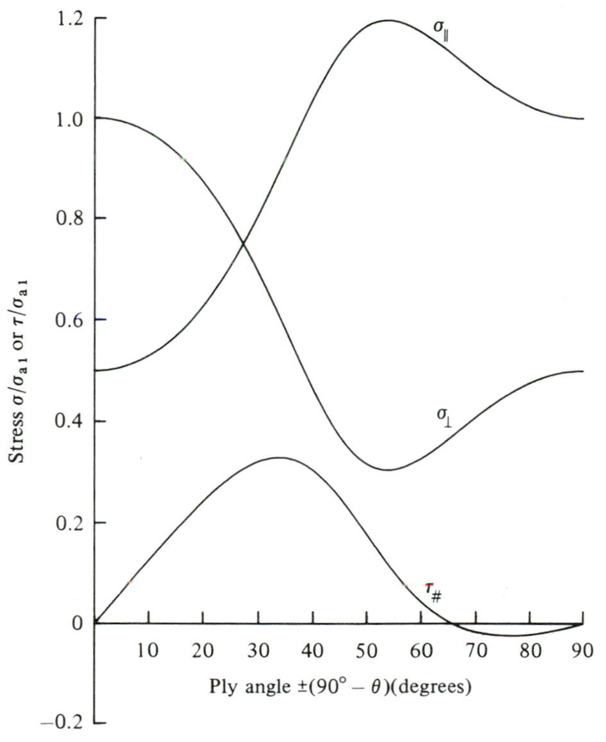

θ and at some angles $\sigma_{\perp}$ becomes compressive. The position of the curves will depend on $E_{\parallel}$, $E_{\perp}$, $\nu_{\parallel\perp}$ and $G_{\#}$ which in turn depend on V_f and the properties of the fibres and resin. The effect of different stress states can be appreciated by reference to Fig. 6.17 which shows the stresses in the laminae in angle-ply laminates tested in biaxial tension with $\sigma_{a1} = 2\sigma_{a2}$.

References and further reading

Ashton, J. E., Halpin, J. C. & Petit, P. H. (1969) *Primer on Composite Materials: Analysis*. Technomic, Westport, Conn.

Chamis, C. C. (1980) Prediction of fiber composite mechanical behaviour made simple. *Proceedings of the 35th SPI/RP Annual Technology Conference*, paper 12-A. Society of the Plastics Industry, New York.

Greenwood, J. H. (1977) German work on G.R.P. design. *Composites* **8**, 175–84.

Jones, R. M. (1975) *Mechanics of Composite Materials*. McGraw-Hill, New York.

Sinclair, J. H. & Chamis, C. C. (1979) Fracture modes in off-axis fiber composites. *Proceedings of the 34th SPI/RP Annual Technology Conference*, paper 22-A. Society of the Plastics Industry, New York.

Spencer, B. & Hull, D. (1978) Effect of winding angle on the failure of filament wound pipe. *Composites* **9**, 263–71.

7 Strength of unidirectional laminae

7.1 Introduction

There are very few, if any, practical applications for composite materials based entirely on unidirectional laminae. This is primarily because the transverse tensile and shear stiffnesses and, in particular, the strengths are much smaller than the corresponding stiffnesses and strengths parallel to the fibres. The former properties are dominated by the matrix properties whereas the latter are dominated by the fibre properties. It was shown in the previous chapter that high stiffness in more than one direction is achieved by laminating the unidirectional layers in predetermined directions to produce the desired properties. Laminates must also be designed to resist fracture. In Chapter 8 the failure of laminates is described in terms of the sequential failure of individual laminae and the interlaminar failure processes. In this chapter the failure processes and the strength of individual laminae are considered.

A lamina can fail in many different modes depending on the external loading conditions. For the purpose of design it is probably sufficient to know the fracture strengths associated with these failure modes for different fibre–resin systems with a range of fibre volume fractions. Some typical values of the fracture strengths for the three main fibres in polyester or epoxy resin are given in Table 7.1. σ^* is the uniaxial strength in tension (T) or compression (C), parallel or perpendicular to the fibre direction. $\tau^*_{\#}$ is the intralaminar shear strength parallel to the fibres. Further definition of these terms is included in the appropriate sections of this chapter.

For the prediction of the strength of laminae in terms of the strength of the fibres and the matrix an understanding of the failure mechanisms is essential. In the first part of this chapter the longitudinal tensile strength is discussed. Initially it is assumed that all the fibres in the lamina have the same strength and then the more realistic situation of variable strength between fibres and a distribution of flaw sizes along each fibre is considered. This leads on to an account of the mechanisms of crack growth and an analysis of the energy required to propagate cracks and its relation to the strength of the fibre–matrix interface. The next section deals with the prediction of transverse strength and the micromechanisms associated with the nucleation and growth of cracks

Table 7.1. *Typical values of strength properties of unidirectional laminae*, $V_f \approx 0.50$

Material	$\sigma^*_{\parallel T}$ (MN m^{-2})	$\sigma^*_{\parallel C}$ (MN m^{-2})	$\sigma^*_{\perp T}$ (MN m^{-2})	$\sigma^*_{\perp C}$ (MN m^{-2})	$\tau^*_{\#}$ (MN m^{-2})
Glass–polyester	650–750	600–900	20–25	90–120	45–60
Type I carbon–epoxy	850–1100	700–900	35–40	130–190	60–75
Kevlar 49–epoxy	1100–1250	240–290	20–30	110–140	40–60

in transverse tension. This is followed by an account of the effect, on the fracture modes and fracture strengths, of testing at angles between the transverse and longitudinal directions. Similarly, the failure processes in transverse and longitudinal compression and in shear are described. The last section deals with the prediction of failure in multiaxial testing conditions. Various failure criteria which take account of the interaction between the applied tensile, compressive and shear stresses, are described.

7.2 Longitudinal tensile strength

Uniform strength fibres

In Section 5.2 it was shown that provided there is a strong fibre–matrix bond $\epsilon_{\parallel} = \epsilon_m = \epsilon_f$ for unidirectional laminae tested in tension parallel to the fibres. The rule of mixtures equation (5.11) was derived using this assumption. Equation (5.11) can be written in the form

$$\sigma_{\parallel} = E_f \epsilon_{\parallel} V_f + E_m \epsilon_{\parallel} (1 - V_f) \tag{7.1}$$

providing the fibre and matrix remain elastic. To a first approximation glass, carbon and Kevlar 49 fibres behave elastically in tension up to the fracture strength σ_f^*. Polyester and epoxy resins have non-linear stress–strain curves and may undergo considerable viscoelastic deformation before fracture as shown in Fig. 2.11. In these circumstances, it is more appropriate to express the stress on the laminae parallel to the fibres as

$$\sigma_{\parallel} = \sigma_f V_f + \sigma_m (1 - V_f) \tag{7.2}$$

where σ_f and σ_m are the stresses on the fibre and matrix respectively.

The response of the composite depends on the relative strains to failure of the matrix and the fibre and two possible situations are considered. Firstly, $\epsilon_f^* > \epsilon_m^*$, and secondly $\epsilon_m^* > \epsilon_f^*$, where ϵ_f^* and ϵ_m^* are the failure strains in uniaxial tension of the fibre and matrix respectively. It is assumed that the strain in the matrix in a unidirectional lamina tested in longitudinal tension is a simple tensile strain so that the failure strain of the matrix in the lamina is the same as in the resin tested alone. This may not be a good assumption, particularly in laminae with high V_f.

When $\epsilon_f^* > \epsilon_m^*$ as illustrated in Fig. 7.1 two different failure sequences can be envisaged depending on V_f. For low V_f the strength of the lamina $\sigma_{\parallel}^*$ depends primarily on σ_m^*. The matrix fractures before the fibre and then all the load is transferred to the fibres as illustrated schematically in Fig. 7.2. When V_f is small, the fibres are unable to support this load and break, thus,

$$\sigma_{\parallel}^* = \sigma_f' V_f + \sigma_m^* (1 - V_f) \tag{7.3}$$

When V_f is large the matrix takes only a small proportion of the load because $E_f > E_m$ so that when the matrix fractures the transfer of load to the fibres is insufficient to cause fracture. Provided it is still possible to transfer the load to the fibres the load on the lamina can be increased until the fracture strength of the fibres is reached, then,

$$\sigma_{\parallel}^* = \sigma_f^* V_f \tag{7.4}$$

The fracture strength varies with V_f as illustrated in Fig. 7.1*b*. The

Fig. 7.1. (*a*) Stress–strain curves of fibre and resin for $\epsilon_f^* > \epsilon_m^*$. (*b*) Variation of fracture strength of unidirectional laminae with V_f for $\epsilon_f^* > \epsilon_m^*$.

(*a*)

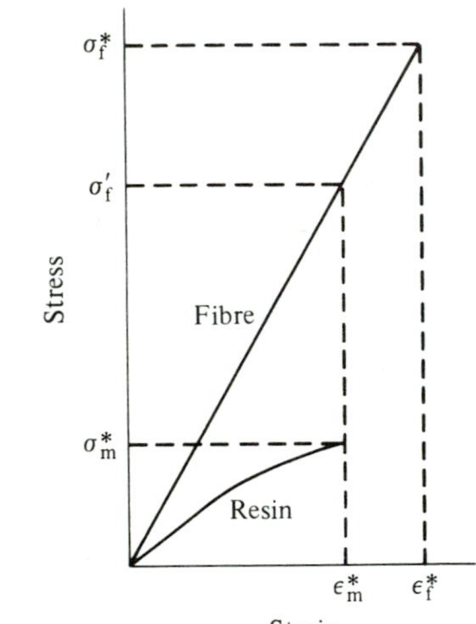

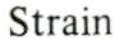

(*b*)

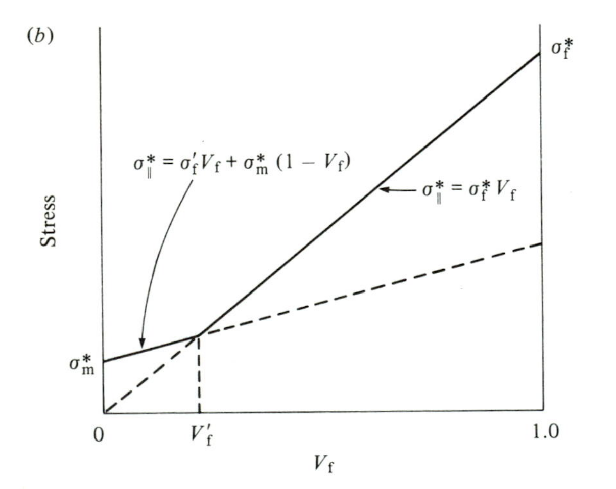

cross-over point is obtained by combining equations (7.3) and (7.4) giving

$$V_f' = \sigma_m^*/(\sigma_f^* - \sigma_f' + \sigma_m^*) \quad (7.5)$$

In a glass fibre–polyester resin lamina with $\epsilon_m^* = 0.020$ and $\epsilon_f^* = 0.025$ the above analysis applies. Taking the following experimentally determined values $\sigma_m^* = 72$ MN m^{-2}, $\sigma_f^* = 2100$ MN m^{-2} and $E_f = 76$ GN m^{-2} gives $\sigma_f' = 1520$ MN m^{-2} and $V_f' = 0.11$. It follows that for all values of V_f above 0.11 the fibre strength dominates. Most commercial laminates have V_f in the range 0.4–0.7.

For $V_f > V_f'$ multiple resin cracking occurs as described by Aveston & Kelly (1973). The phenomenon is important in a number of aspects of the fracture of composite materials and is described in more detail in Section 8.2 in connection with the failure processes in laminates. Multiple cracking is illustrated in Fig. 7.2*c* and arises because the first resin cracks that are evident in Fig. 7.2*b* do not result in complete unloading of the resin since the fibre and resin are bonded together. As the strain on the lamina increases, further resin cracking occurs. The final spacing of the cracks depends on the ratio E_f/E_m, the strength of the bond and the difference in the failure strains of resin and fibre.

The same approach can be applied to laminae in which $\epsilon_m^* > \epsilon_f^*$ and again two different failure sequences can occur depending on V_f (Fig. 7.3). When fibre fracture occurs in low V_f laminae (Fig 7.2*d*) the extra load on the matrix is not sufficient to fracture the matrix. However, since the effective cross-section of the matrix is reduced by the presence of the 'holes' at the fibre ends the load carrying capacity is less than σ_m^*

Fig. 7.2. Schematic representation of longitudinal tensile failure of a unidirectional lamina (*a*) before fracture, (*b*) resin fracture before fibre fracture, $\epsilon_f^* > \epsilon_m^*$, (*c*) multiple resin cracking, (*d*) fibre fracture before resin fracture, $\epsilon_m^* > \epsilon_f^*$.

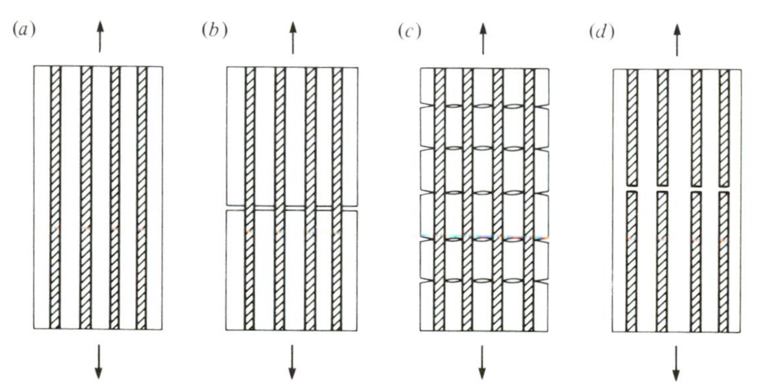

by an amount proportional to V_f, and,

$$\sigma_{\parallel}^* = \sigma_m^* V_m = \sigma_m^* (1 - V_f) \tag{7.6}$$

Clearly, when V_f is large the load transferred to the matrix when fibre fracture occurs is very large, and cannot be supported, so that the matrix fractures when the fibres fracture, and

$$\sigma_{\parallel}^* = \sigma_f^* V_f + \sigma_m'(1 - V_f) \tag{7.7}$$

The variation of $\sigma_{\parallel}^*$ with V_f is illustrated in Fig. 7.3*b* and the cross-over point obtained from equations (7.6) and (7.7) is

$$V_f' = (\sigma_m^* - \sigma_m')/(\sigma_f^* + \sigma_m^* - \sigma_m') \tag{7.8}$$

Fig. 7.3. (*a*) Stress–strain curves of fibre and resin for $\epsilon_m^* > \epsilon_f^*$. (*b*) Variation of fracture strength of unidirectional laminae with V_f for $\epsilon_m^* > \epsilon_f^*$.

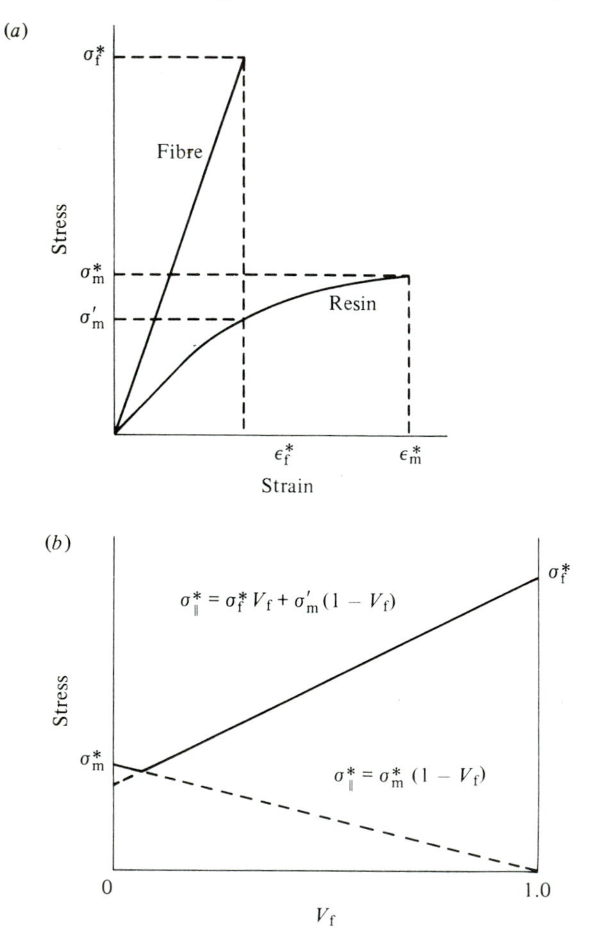

This analysis can be applied to carbon fibre–epoxy resin systems since the strain to failure of the fibres is less than that of the resin. Taking some typical values for a Type I carbon fibre and an epoxy resin, i.e. $\epsilon_f^* = 0.005$, $\epsilon_m^* = 0.02$, $\sigma_m^* = 80$ MN m^{-2}, $E_m = 5.3$ GN m^{-2}, $\sigma_f^* = 2.0$ GN m^{-2} gives $\sigma_m' = 26.5$ MN m^{-2} and $V_f' = 0.026$. Similarly, for glass fibres in a fairly flexible polyester resin with experimental values $\epsilon_f^* = 0.025$, $\epsilon_m^* = 0.035$, $\sigma_m^* = 65$ MN m^{-2}, $\sigma_m' = 52$ MN m^{-2}, $\sigma_f^* = 2.1$ GN m^{-2} gives $V_f' = 0.006$. In both these examples V_f' is very small indicating that the matrix contribution to the longitudinal tensile strength of polymer-based fibre composite materials is small and can be neglected. Equation (7.4) then applies.

Variable strength fibres

The above treatment neglects some significant features of the strength of fibres such as the variation of strength from fibre to fibre and the distribution of flaw sizes along each fibre. Some of the experimental evidence for the variation of strength has already been given in Chapter 2. Reference to Fig. 2.8 indicates that the fibre strength

Fig. 7.4. Weibull cumulative probability distribution function $G(\sigma)$.

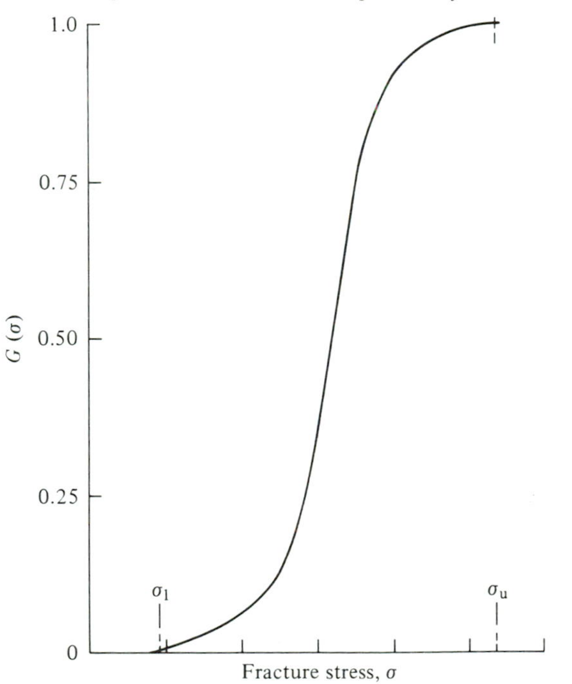

is not a unique specific value; it varies from one fibre to another and depends on the length over which it is measured. Thus, the choice of σ_f^* in equation (7.4) is far from obvious and it is necessary to use a statistical approach (see Corton 1967 for a review).

It is well known that the experimentally measured strengths of brittle solids are subject to large amounts of scatter which can be described by the *Weibull distribution function*. This approach can also be used for brittle fibres. When the strength of a large number of fibres, diameter $2r$, is measured over a test gauge length l the distribution of strengths can be represented by the Weibull distribution function. It may be assumed, in order to provide a physical insight into the analysis, that the Weibull distribution curve has two cut-off points corresponding to a lower strength limit σ_l and an upper strength limit σ_u. The latter is equivalent to the theoretical strength of the fibre in the absence of any internal defects or surface flaws which are responsible for the reduced strengths measured. Fig. 7.4 shows the Weibull cumulative probability distribution function $G(\sigma)$ which is given by

$$G(\sigma) = 1 - \left[1 - \left(\frac{\sigma - \sigma_l}{\sigma_o}\right)^m\right]^\omega \tag{7.9}$$

and is the probability of fracture of a fibre at a stress level equal to or less than σ. In this equation $\sigma_o = \sigma_u - \sigma_l$, ω is related to the dimensions of the test sample and for fibres with a constant diameter is given by $l/2r$, and m depends on the amount of scatter. It is related to the statistical distribution parameters by

$$s/\bar{\sigma} \approx 1.2/m \tag{7.10}$$

where $s/\bar{\sigma}$ is the coefficient of variation in which s, the standard deviation, is given by

$$s = \left[\frac{\sum_{i=1}^{N} (\sigma_i - \bar{\sigma})^2}{N}\right]^{\frac{1}{2}} \tag{7.11}$$

and $\bar{\sigma}$, the mean strength, is given by

$$\bar{\sigma} = \sum_{i=1}^{N} \sigma_i \Big/ N \tag{7.12}$$

The fracture strength of a bundle of fibres σ_b which, as individual fibres, exhibit the distribution described by equation (7.9), and have a mean strength $\bar{\sigma}$, was derived by Coleman (1958). It is envisaged that the fibres in the bundle are separated from each other and have the same cross-sectional area. At all stresses below σ_l the fibres have the same elongation and are unbroken. As the load is increased the weakest fibres break in succession and the load must be transferred to the unbroken fibres. The maximum or breaking load of the bundle occurs when the stress on the remaining fibres reaches σ_u and complete fracture occurs. It is found that σ_b is less than $\bar{\sigma}$ and that the ratio $\sigma_b/\bar{\sigma}$ depends on the amount of scatter in the strength of the individual fibres according to

$$\frac{\sigma_b}{\bar{\sigma}} = \left(\frac{1}{me}\right)^{1/m} \frac{1}{\Gamma(1+1/m)} \tag{7.13}$$

where Γ is a tabulated gamma function. The relation between $\sigma_b/\bar{\sigma}$ and m is shown graphically in Fig. 7.5. Thus, for very brittle fibres which usually show a large amount of scatter and have values of m between 2 and 5, the bundle strength is 50–65% of the mean strength. Typically, glass fibres have values of m between 5 and 15 so that the bundle strength is 65–80% of the mean strength. When m is infinite, i.e. the coefficient of variation is zero, all the fibres have the same strength and the bundle strength is equal to the individual fibre strength. This assumption is, of course, made in deriving the rule of mixture equations (7.3) and (7.7).

The analysis of bundle strength is very useful for determining a value for fibre strength but it is not an accurate representation of failure when the fibres are bonded together by resin in the composite material. Two other models have to be considered, one due to Rosen called the *cumulative weakening model*, and the other due to Zweben called the *fibre break propagation model*.

The basis of the *cumulative weakening model* is shown schematically in Fig. 7.6. It is assumed that flaws in the fibres are distributed randomly and that the statistical information for the strength of single fibres applies. When the lamina is loaded the stress on the fibres is related to the stress on the lamina by equation (7.2). As the load increases, the

Fig. 7.5. Effect of scatter in strength (m) on the ratio of bundle strength to average strength.

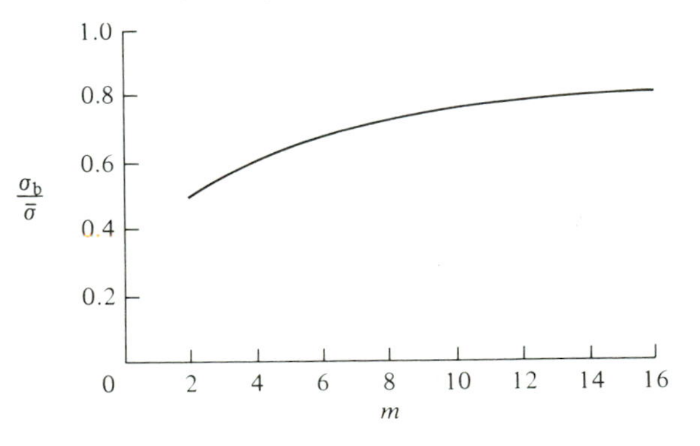

fibres fail at their weakest points. One possibility is that the first fracture will immediately propagate across the whole section (the weakest link failure model). This only occurs when the fibres are very strongly bonded to the matrix and the matrix is very brittle. In general, as will be shown later, the single fibre cracks do not propagate immediately. Following the arguments in Section 5.4, the strains around the broken ends of the fibre are modified and there is a length of the fibre which does not support the full load, i.e. the ineffective length l_c. The volume of material affected in this way depends on the interfacial shear strength. Progressive fibre fracture produces cumulative weakening and redistribution of load. For this model, it is proposed that failure occurs when a layer across the section of the lamina is so weakened that it can no longer sustain the applied load. The statistical treatment due to Rosen predicts that

$$\frac{\sigma^*_{\text{cum}}}{\bar{\sigma}} = \left(\frac{l}{l_c m e}\right)^{1/m} \frac{1}{\Gamma(1+1/m)} \tag{7.14}$$

where σ^*_{cum} is the fibre strength expected in the composite and l_c is the ineffective length. The difficulty in relating this expression to experimental data is in selecting a value for l_c. However, it is likely to be in the range 10–100 fibre diameters for most fibres so that, for a test sample gauge length at 100 mm, $l/l_c \approx 10^4$. Taking $l/l_c = 10^4$ and $m = 5$ and 10 gives $\sigma^*_{\text{cum}}/\bar{\sigma} = 4.07$ and 1.96 respectively. The corresponding values for

Fig. 7.6. Illustration of Rosen's cumulative weakening model of lamina fracture.

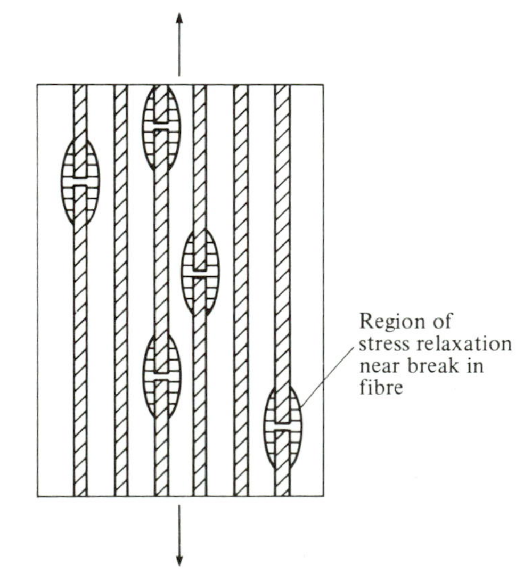

$l/l_c = 10^2$ are 1.62 and 1.24 respectively. Thus, in contrast to σ_b (Fig. 7.5) σ^*_{cum} exceeds $\bar{\sigma}$ so that the predicted strength of the composite $\sigma^*_{cum}\ V_f$ is greater than that predicted on the basis of mean fibre strength. Clearly, there will be an upper limit to the lamina strength defined by $\sigma_u\ V_f$.

The *fibre break propagation model* is probably more realistic in terms of the actual failure process but it is very difficult to obtain a meaningful estimate of the strength. The initial stages of fracture follow those described above for the cumulative weakening model. As each fibre breaks, the redistribution of stress leads to additional stresses on neighbouring fibres, i.e. there is a stress magnification effect. Thus, there is an increased probability that fracture will occur in the immediately adjacent fibres. This increases as the load increases and eventually sequential fibre fracture occurs. The model is illustrated schematically in Fig. 7.7. The lamina strength predicted from the model will be lower than $\sigma^*_{cum}\ V_f$ since the cumulative weakening model neglects the stress concentration effects around broken fibres. It will also depend on the microprocesses of deformation and fracture which occur at the broken ends since they determine the magnitude of the stress magnification and the volume over which the stress is redistributed.

Interaction between cracks and fibres

Consider the broken fibre illustrated in Fig. 7.8 which is embedded in a matrix and surrounded by a set of parallel fibres. The analysis in Section 5.4 shows that for a fibre perfectly bonded to the

Fig. 7.7. Fracture in adjacent fibres due to stress concentration at tip of first fibre fracture.

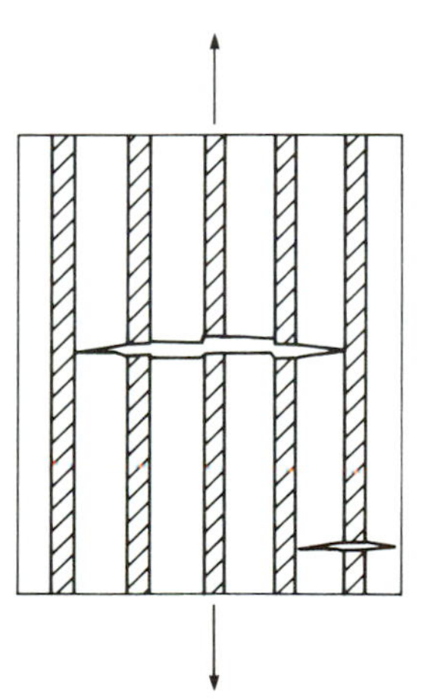

matrix the shear stress concentration parallel to the interface is a maximum at the end of the fibre. If τ_{max} exceeds the shear strength of the interface, or the surrounding matrix material, fracture will occur at or close to the interface. The fracture will start at the fibre end and grow along the fibre as the stress on the system increases. The three main processes which can occur at the broken fibre are illustrated in Fig. 7.8: (*a*) the crack in the fibre has propagated into the surrounding matrix as a brittle crack, (*b*) the matrix has yielded so blunting the sharp crack and the yield zone has spread along the fibre, and (*c*) the interface or the matrix immediately adjacent to the fibre or the fibre immediately adjacent to the matrix has failed in shear allowing the unloaded fibre to shrink back into the matrix. Shear yielding or shear failure does not necessarily imply that the matrix is unable to transfer load to the fibre and in Section 7.3 it is shown that frictional forces at the interface lead to a linear build-up in stress in the fibre from the broken ends. The relative importance of each of these processes depends on the properties of the fibres and the resin and on V_f.

Some of the important interactions between failure processes and material parameters can be understood by considering the processes which occur when the brittle, sharp crack shown in Fig. 7.8*a* meets the next fibre. The stress concentration around a crack is proportional to $(c/\rho)^{\frac{1}{2}}$, where ρ is the radius of curvature at the crack tip and $2c$ is the length of the crack. For a crack subjected to a uniaxial tensile stress normal to the plane of the crack the presence of the crack results in additional tensile stresses parallel to the plane of the crack. The stress system close to the tip of an elliptical crack in an isotropic material is shown in Fig. 7.9. The maximum tensile stress $\sigma_{1\,max}$ at right angles to the crack occurs at the tip of the crack and the maximum tensile stress $\sigma_{2\,max}$ parallel to the crack occurs just ahead of the crack tip. For isotropic materials, the ratio $\sigma_{1\,max}/\sigma_{2\,max}$ is approximately 5. There is also

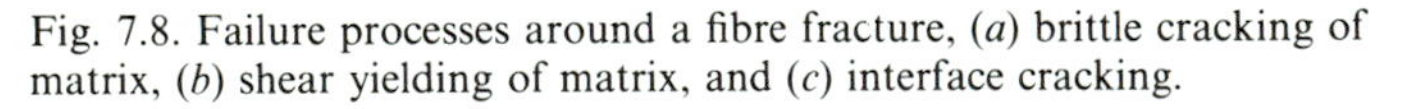
Fig. 7.8. Failure processes around a fibre fracture, (*a*) brittle cracking of matrix, (*b*) shear yielding of matrix, and (*c*) interface cracking.

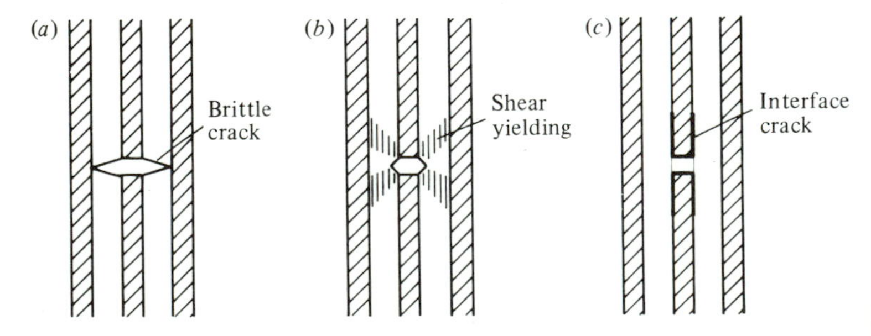

a shear stress τ in the plane normal to the plane of the crack and $\sigma_{1\,\max}/\tau_{\max} \approx 3.4$. In anisotropic materials $\sigma_{1\,\max}/\sigma_{2\,\max}$ and $\sigma_{1\,\max}/\tau_{\max}$ will depend on the orientation of the crack and the degree of anisotropy. For a crack oriented as shown in Fig. 7.8 Kelly (1970) quotes $\sigma_{1\,\max}/\sigma_{2\,\max} = 48$, $\sigma_{1\,\max}/\tau_{\max} = 11$ and $\tau_{\max}/\sigma_{2\,\max} = 4.4$ using data applicable to a carbon fibre–epoxy resin lamina with $V_f = 0.5$.

From Fig. 7.9 it can be seen that when a crack in the matrix meets a fibre the stress σ_1 at the tip of the crack will tend to cause fibre fracture, the stress σ_2 will lead to tensile separation at the interface and the stress

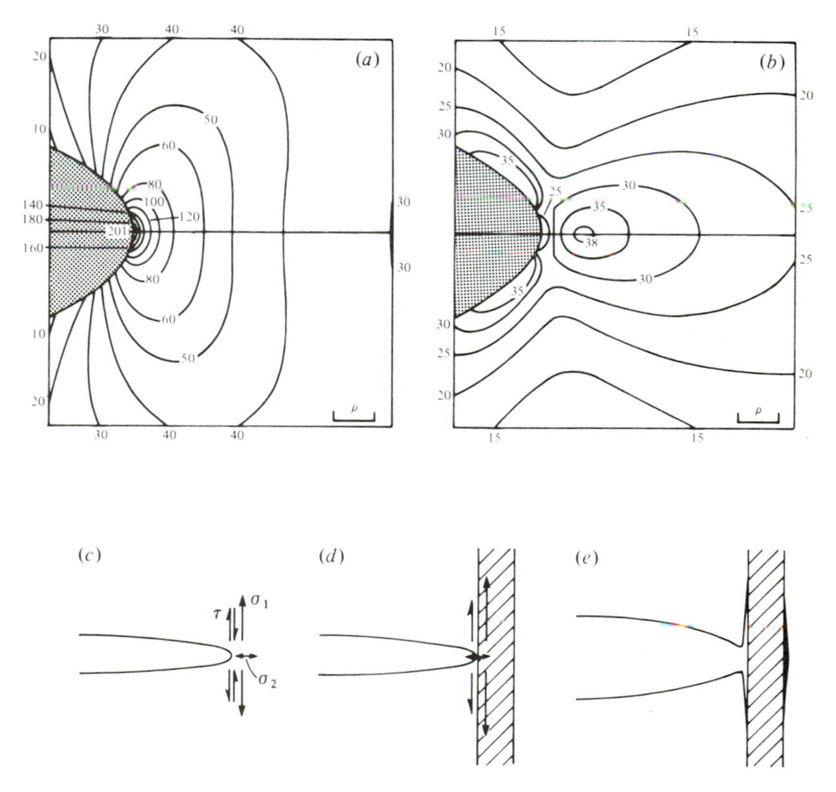

Fig. 7.9. Stress system close to an elliptical crack in an elastic solid meeting a fibre interface. (After Cook and Gordon 1964.) (*a*) Stress concentrations produced at right angles to the plane of a crack parallel to the applied load. Figures indicate the number of times by which local stress is increased as compared with mean stress remote from crack (σ_1). (*b*) As for (*a*) stress concentrations parallel to the plane of crack, at right angles to the applied load (σ_2). (*c*) Schematic representation of σ_1, σ_2 and τ at crack tip. (*d*) Crack tip at fibre interface. (*e*) Interface splitting and crack opening when crack intersects fibre.

τ will cause shear failure at the interface. The processes which occur depend on the critical stresses for these failure processes. By way of example, suppose that these critical stresses are given by $\sigma_\parallel^*$, $\sigma_\perp^*$ and $\tau_\#^*$. When $\sigma_\parallel^*/\sigma_\perp^* > \sigma_{1\,\max}/\sigma_{2\,\max}$ tensile cracking parallel to the interface will occur before fibre fracture, when $\sigma_\parallel^*/\tau_\#^* > \sigma_{1\,\max}/\tau_{\max}$ shear fracture will occur before fibre fracture and when $\tau_\#^*/\sigma_\perp^* > \tau_{\max}/\sigma_{2\,\max}$ tensile cracking at the interface will be preferred to shear cracking. The values of $\sigma_\parallel^*$, $\sigma_\perp^*$ and $\tau_\#^*$ given in Table 7.1 indicate that there are significant differences in the critical stress ratios from one system to another. It is important to recognise that these ratios are affected by factors such as resin and interface chemistry, surface treatment of fibres, volume fraction of fibres and environmental conditions such as temperature and humidity. All the systems in Table 7.1 give $\tau_\#^*/\sigma_\perp^*$ between 2 and 3 which suggests, following the argument in the preceding paragraph, that shear cracking is more likely than tensile cracking at the interface. Similarly, $\sigma_\parallel^*/\tau_\#^*$ is between 12 and 45 so that shear cracking at the interface will occur before fibre fracture. The stress

Fig. 7.10. Scanning electron micrograph of fracture surface of a carbon fibre–epoxy resin lamina tested in longitudinal tension. Fracture surface is relatively smooth and consists of a network of blocky outcrops of fibres and resin at different levels. (From J. D. Grundy, unpublished data.)

concentration in the fibre is relaxed by interface cracking but complete unloading does not occur, particularly with shear cracking, since frictional forces still operate. The stress in the fibres can build up and cause failure.

The relative amounts of interface failure and the magnitude of the frictional forces determine the overall appearance of the fracture surface. For a strongly bonded carbon fibre–epoxy resin system with a relatively small value of $\sigma_{\parallel}^{*}/\tau_{\#}^{*}$ there is only a small amount of debonding and the fracture surface is fairly smooth as shown in Fig. 7.10. In contrast Figs. 7.11 and 7.12 show the fracture surfaces of glass fibre–polyester resin and Kevlar 49 fibre–epoxy resin systems respectively which have

Fig. 7.11. Scanning electron micrograph of fracture surface of a glass fibre–polyester resin lamina tested in longitudinal tension illustrating brush-like appearance associated with extensive fibre pull-out.

Fig. 7.12. Scanning electron micrographs of fracture surface of a Kevlar 49–epoxy resin lamina tested in longitudinal tension. (*a*) Fibrous fracture showing necking of individual fibres. (*b*) High magnification view of fibre X in (*a*) showing fibre fibrillation and kink band formation within fibrils at Y.

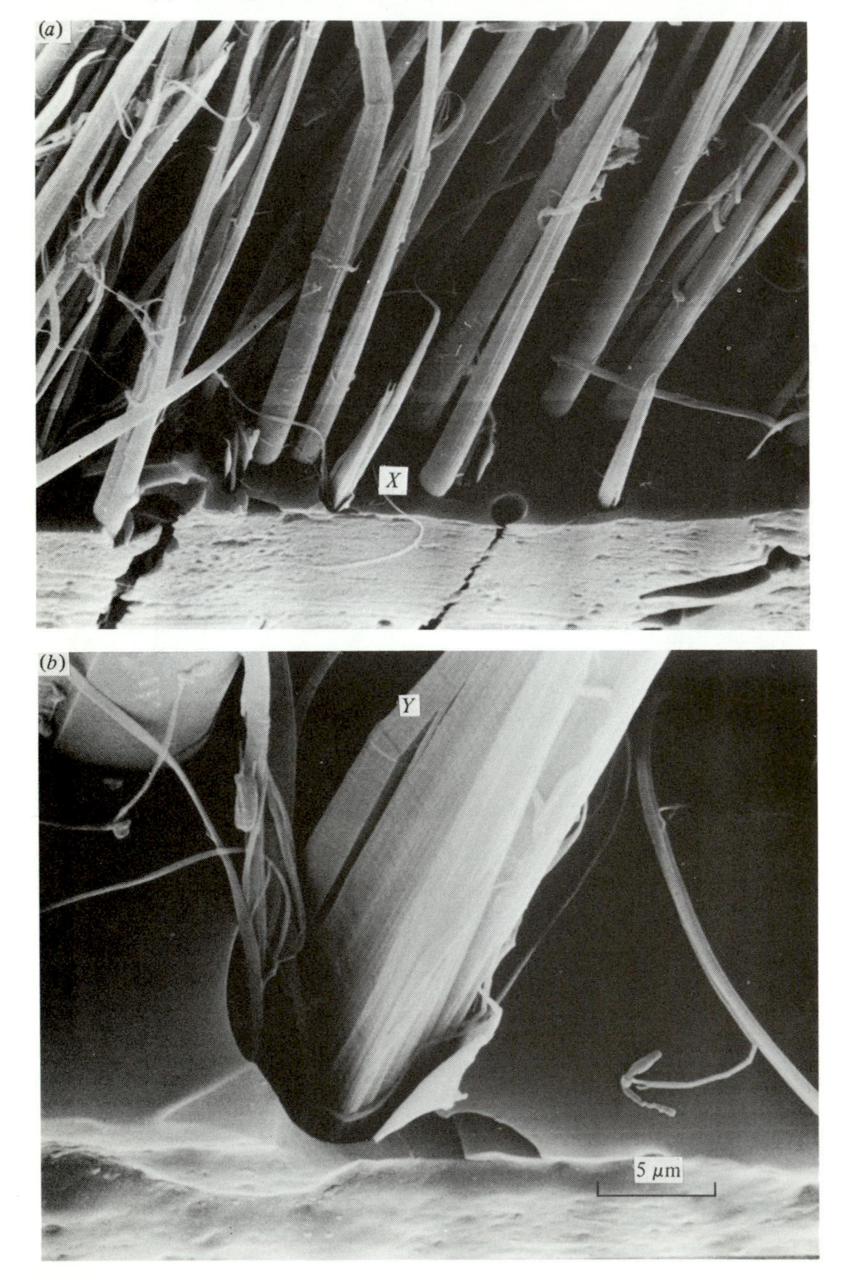

relatively high values of $\sigma_{\parallel}^{*}/\tau_{\#}^{*}$ and $\sigma_{\parallel}^{*}/\sigma_{\perp}^{*}$ and undergo massive debonding. The fracture surface is very fibrous in appearance with large amounts of fibre pull-out. A dramatic change in the glass fibre–polyester resin fracture surface appearance occurs when the material is tested in

Fig. 7.13. Scanning electron micrograph of fracture surface of a glass fibre–polyester resin lamina tested in longitudinal tension in presence of dilute hydrochloric acid. (From P. J. Hogg, PhD thesis, University of Liverpool 1981.)

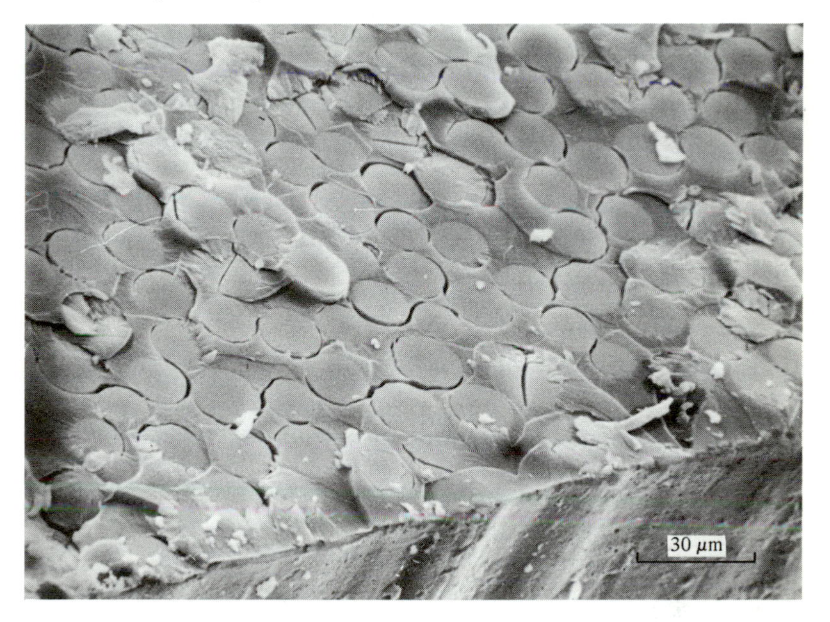

Fig. 7.14. Schematic representation of fracture surfaces in Fig. 7.10–7.13 (*a*) fibrous surface due to debonding and fibre pull-out (*b*) planar fracture.

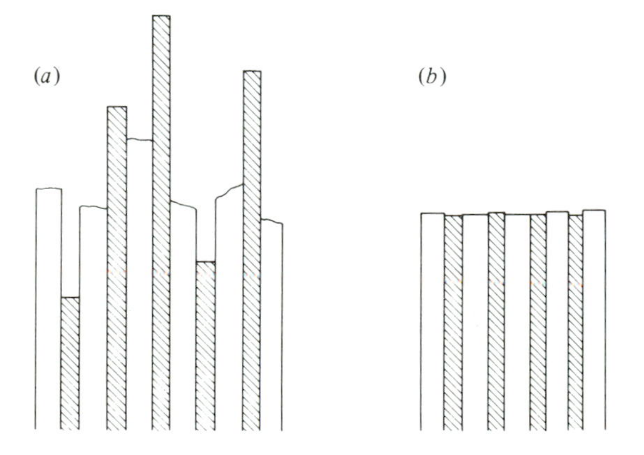

a mineral acid, such as hydrochloric acid. Very smooth fracture surfaces (Fig. 7.13) arise because the acid has the effect of producing a large reduction in $\sigma_{\parallel}^*$ and hence $\sigma_{\parallel}^*/\tau_{\#}^*$. Schematic representations of these fracture surfaces are shown in Fig. 7.14.

7.3 Fibre pull-out

It will be evident from Fig. 7.6 that cracking of fibres and resin can occur before complete separation of the fracture surfaces. The two surfaces are held together by fibres which bridge the fracture plane. In this section, which is based on some important fundamental studies by Kelly and his co-workers reported in Kelly (1970) and by Outwater & Murphy (1969), it is shown that the work done in separating the surfaces makes a major contribution to the total energy of fracture. The fracture toughness of composite materials is discussed in Section 10.3.

The basic principles can be understood by reference to the single fibre experiment illustrated in Fig. 7.15. The fibre is embedded in the matrix over the length l_e. If the shear strength of the interface is τ, the tensile stress on the fibre required to produce bond breakage is determined to a first approximation by balancing the tensile and shear stresses, thus

$$\sigma \pi r^2 = 2\pi r l_e \tau \quad \text{or} \quad l_e/r = \sigma/2\tau^* \tag{7.15}$$

Fig. 7.15. Pull-out test (*a*) fibre embedded in resin, (*b*) typical variation of pull-out stress with embedded length.

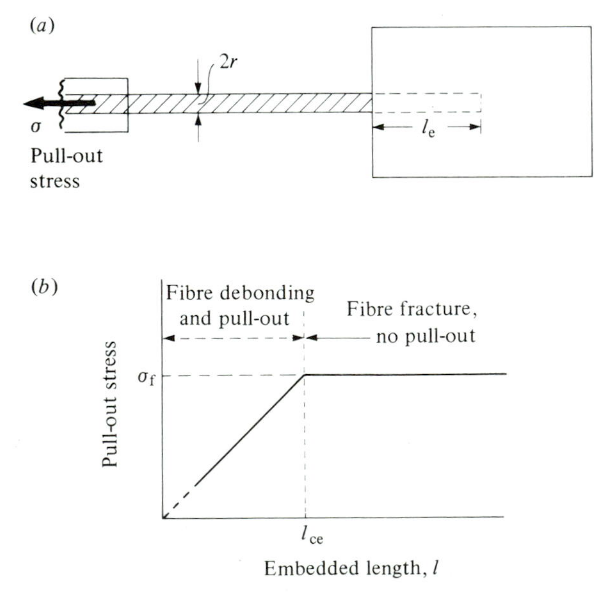

When the tensile stress required for bond breakage is plotted against the embedded length, l_e, there is a sharp cut-off due to fibre fracture before debonding as shown in Fig. 7.15*b*. The critical fibre length l_{ce} depends on the strength of the bond. This technique can be used to measure τ (see Section 3.5) since

$$l_{ce} = \sigma_f^* r/2\tau \tag{7.16}$$

where σ_f^* is the fracture strength of the fibre.

Experimentally, it is found that for fibres embedded in brittle polyester and epoxy matrices, the stress required to extract the fibre does not drop to zero after debonding has occurred because there are large frictional forces which resist the sliding of the fibre out of the resin sheath. This is best illustrated in a slightly different experiment as shown in Fig. 7.16. A short length of the fibre is bonded to a disc of the resin and the load required to pull the fibre through the disc is measured. The load–displacement curve shows that there is a peak associated with debonding and that after debonding an approximately constant load is required. The frictional forces are usually attributed to residual stresses associated with resin shrinkage during curing and to differential thermal contraction. In addition the reduction in the stress on the fibre

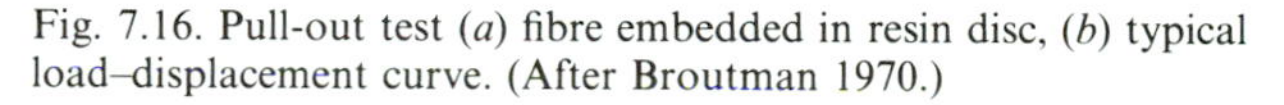

Fig. 7.16. Pull-out test (*a*) fibre embedded in resin disc, (*b*) typical load–displacement curve. (After Broutman 1970.)

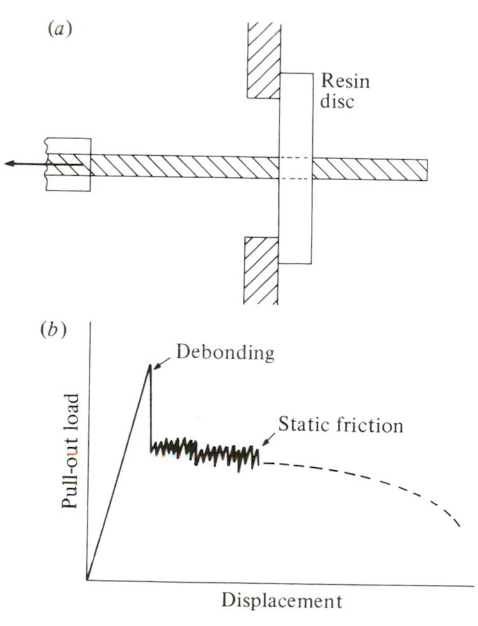

at debonding results in an increase in fibre diameter owing to the Poisson expansion and this leads to an increase in the pressure on the fibre surface. The relative contribution of the interface bonding and frictional forces to the work required to pull the fibre out of the matrix depends on a large number of material parameters. However, it is clear from Fig. 7.16 that the frictional forces could make a major contribution if the distance over which pull-out occurs is large.

Before making an estimate of the work done in fibre debonding and fibre pull-out, reference must be made to the stress distribution along the fibre after debonding has occurred. Consider the fibre loaded as illustrated in Fig. 5.10. The analysis in Section 5.4 shows that debonding will initiate at the ends of the fibre and then spread along the fibre. The frictional forces will still operate and allow the transfer of stress to the fibre. If the friction force is constant, the load on the fibre will increase linearly from the ends of the fibre according to

$$\pi r^2 \mathrm{d}\sigma = 2\pi r\tau\,\mathrm{d}x \quad \text{or} \quad \mathrm{d}\sigma/\mathrm{d}x = 2\tau/r \tag{7.17}$$

where τ is the frictional stress at the interface. The rate of increase is lower than that in the unfailed elastic case (Fig. 5.10) as illustrated schematically in Fig. 7.17. The plateau corresponds to the stress level in a continuous, unbroken fibre under the same loading conditions and l_c is the ineffective length. For short fibres of length $l < l_c$ the maximum stress is less than that in a continuous fibre. The reinforcing efficiency is reduced.

The same stress distribution as equation (7.17) applies to a fibre held by frictional forces across a fracture plane as shown in Fig. 7.18 and in the single fibre test illustrated in Fig. 7.15*a*. Clearly, the stress required to extract the fibre increases as the embedded length increases (Fig. 7.15*b*) and the maximum or critical length l_{ce} which can be extracted without fibre fracture is given by equation (7.16).

Fig. 7.17. Tensile stress in fibre due to friction forces.

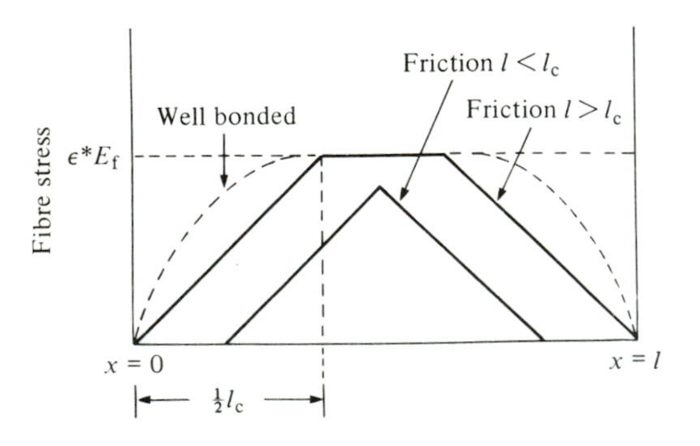

According to Kelly (1970) the work of fracture includes (i) a small contribution from the fracture energy of resin and fibres which are both brittle, (ii) the debonding energy which, for a single fibre, is

$$W_D = \tfrac{1}{6}\pi r^2 (\sigma_f^*/E_f)\,\sigma_f^*\,x \tag{7.18}$$

for debonding over a length x, and (iii) pull-out energy given by

$$W_e = \pi r x^2 \tau \tag{7.19}$$

Since, from equation (7.16), the maximum length which can be extracted is l_{ce} the maximum pull-out energy can be obtained by equating l_{ce} with x to give

$$W_e = \tfrac{1}{2}\pi r^2\,\sigma_f^*\,x \tag{7.20}$$

Thus the ratio of the maximum work done in pulling-out to the work done in debonding is

$$W_e/W_D = 3E_f/\sigma_f^* \tag{7.21}$$

From the ratios E_f/σ_f^* for the fibres listed in Table 2.1, it is clear that pull-out is far more significant than debonding as an energy absorber. However, debonding must occur before pull-out. If the shear stress for debonding is large then fibre fracture occurs before extensive debonding and the amount of pull-out, and hence energy absorbed, is small.

7.4 Transverse tensile strength

The low transverse tensile strength of unidirectional laminae presents a major problem in the design of composite laminate structures. Although the fibres can be oriented so that they are parallel to the external loads, it is almost impossible to avoid transverse stresses which

Fig. 7.18. Broken fibres spanning a resin crack after debonding has occurred.

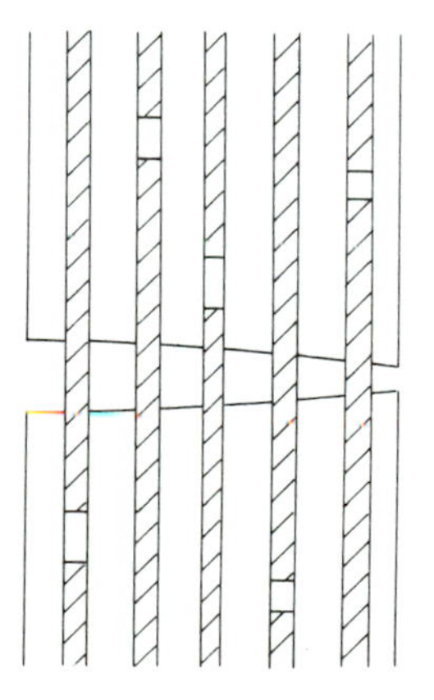

may lead to premature failure of the laminate. An excellent example of this is in the design of filament wound pipe for internal pressure applications which is referred to in Section 8.3.

There is no simple relation for predicting the transverse strength. Unlike the longitudinal tensile strength which is determined almost entirely by a single factor, i.e. the fibre strength (equation (7.4)), the transverse strength is governed by many factors including the properties of the fibre and matrix, the interface bond strength, the presence and distribution of voids, and the internal stress and strain distribution due to the interaction between fibres, voids etc. Some of these aspects have already been introduced in earlier sections. The most clear-cut feature of the transverse strength is that it is usually less than the strength of the parent resin, so that, in contrast to the effect of fibres on the transverse modulus, the fibres have a negative reinforcing effect. In this section the origin and likely extent of the reduction in transverse strength due to the presence of fibres under various conditions is discussed.

Some typical stress–strain curves for unidirectional laminae tested in transverse tension are shown in Fig. 7.19 along with corresponding curves for the matrix material. A small amount of non-linearity has occurred in the laminae with flexible resin matrices indicating that some viscoelastic or plastic flow has occurred. The strains to failure of the laminae appear to be unrelated to those of the matrices and are very small. Similar results have been obtained for many fibre–resin systems (see Table 7.1).

The transverse tensile strength of a lamina in which there is little or no interface bonding is determined by the strength of the resin. To a first approximation the fibres can be regarded as cylindrical holes. For a simple square array of fibres (Fig. 4.2*b*) the cross-sectional area of the matrix is reduced by a factor

$$S/2R = [1 - 2(V_f/\pi)^{\frac{1}{2}}] \tag{7.22}$$

of the original value. Thus the predicted strength, provided that the resin is not notch sensitive is given by

$$\sigma_{\perp}^{*} = \sigma_{m}^{*}[1 - 2(V_f/\pi)^{\frac{1}{2}}] \tag{7.23}$$

The form of this equation is shown in Fig. 7.20. At $V_f = 0.785$ the fibres are touching and the strength falls to zero. A similar relation can be used for other fibre arrangements.

A further reduction in transverse strength is expected if the stress concentrations around the holes are not relieved by plastic flow. For circular holes the maximum stress concentration occurs at the edges of the hole and for a rigid elastic solid has a value of 3. In practice

Fig. 7.19. (*a*) Tensile stress–strain curves for three polyester resins with increasing flexibility and strains to failure. (*b*) Stress–strain curves of unidirectional laminae of glass fibre–polyester resins (same series as in (*a*)), $V_f = 0.48$, tested in transverse tension. *A*, *B* and *C* in (*b*) correspond with *A*, *B* and *C* in (*a*); vertical arrows represent fracture. (From M. J. Legg, PhD thesis, University of Liverpool 1980.) (Note the different scales on the two axes.)

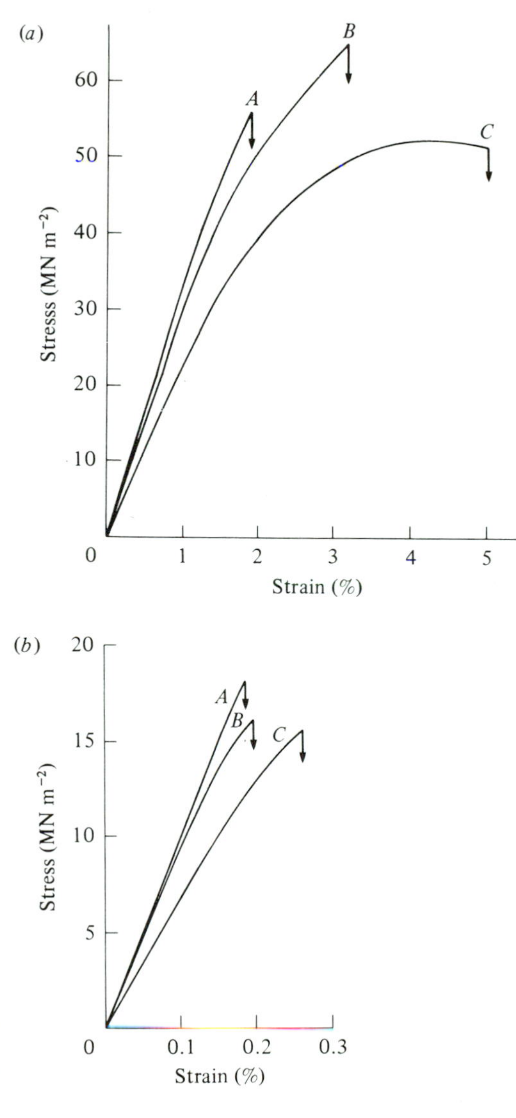

both elastic and plastic deformation will lead to a reduction in this value.

When the fibres are strongly bonded to the matrix the transverse strength is dependent on the strength of the matrix and the interface

Fig. 7.20. Reduction in transverse tensile strength due to presence of cylindrical 'holes' to represent unbonded fibres.

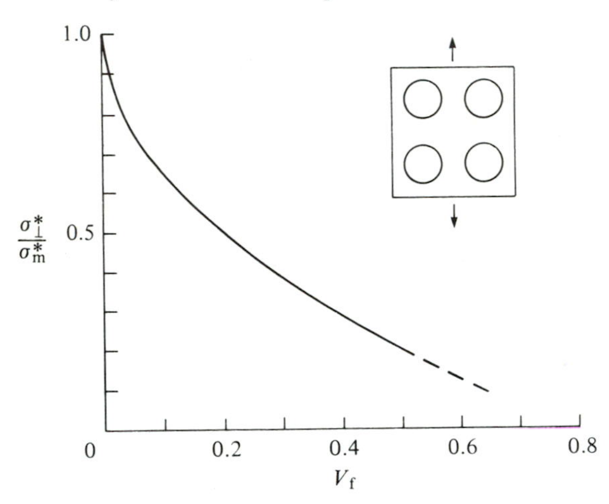

Fig. 7.21. Fracture under transverse loading showing tangential crack in resin matrix around a well-bonded fibre. (After Tirosh *et al.* 1979.)

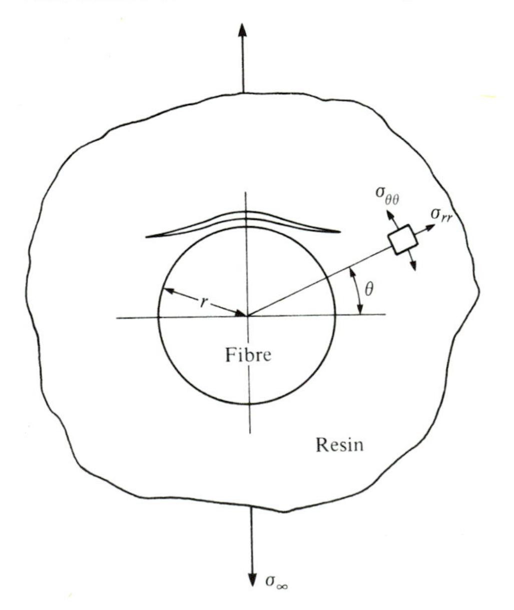

bond. The account in Section 5.2 shows that if the bond does not fail there is a stress and strain magnification in the resin which is a maximum between the fibres. In a theoretical analysis of single fibres embedded in the resin (Fig. 7.21) it was found that the maximum stress magnification occurs along the line $\theta = 90°$ as a small distance $r^* \approx 1.2r$ from the fibre and varies between 1.53 and 1.74 for resins with Poisson's ratio between 0.25 and 0.5. Similarly, calculations of the stress fields between fibres show that the maximum tensile stress concentration occurs midway between the fibres with a value of approximately 2.0 for $V_f \approx 0.5$. Although resin failure will depend on the full triaxial state of stress, it is clear that these stress concentration effects will result in a reduction in strength compared with the resin by a factor of about 2. However, it must be noted that the tensile strength of brittle resins is dominated by the effect of flaws and so the interpretation of the strength reduction in terms of the stress concentration effects of fibres is particularly difficult.

The presence of voids in a well-bonded lamina also leads to stress concentration effects and the discussion above on the strength of laminae with unbonded fibres is relevant. Some results on the effect of voids on transverse strength are shown in Fig. 7.22. The theoretical line is based

Fig. 7.22. Variation of transverse tensile strength of glass fibre–epoxy resin laminae with void content. Theory due to Greszczuk (1974) using properties of constituents, $E_f = 86\,\text{GN m}^{-2}$, $E_m = 3.8\,\text{GN m}^{-2}$, $V_f = 0.55$.

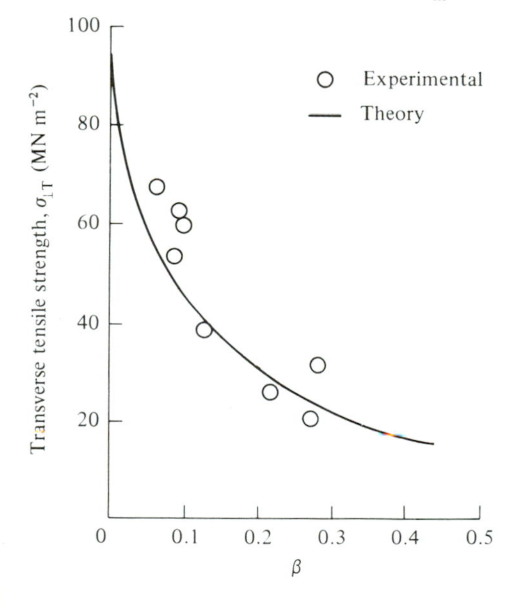

on the combined effects of bonded fibres and voids on the maximum stress concentration and β which is given by

$$\beta = (V_v + V_{if})/[1 - (V_f - V_{if})] \tag{7.25}$$

V_v and V_f have been defined in Chapter 4 (see equation (4.13)) and V_{if} is the volume fraction of fibres made ineffective by the presence of voids.

The stress magnification approach is not appropriate for resins which show a pronounced non-linear stress–strain response. The reduction in transverse strength is then accounted for in terms of the strain magnification which is a maximum between fibres as described in Section 5.2. Very high strain magnification values are obtained when

Fig. 7.23. Propagation of a transverse crack in a glass fibre–polyester resin lamina. Tensile strain to failure of pure resin was 1.6% and glass fibres were coated with a resin compatible silane coupling agent. (From M. L. C. Jones, PhD thesis, University of Liverpool 1981.)

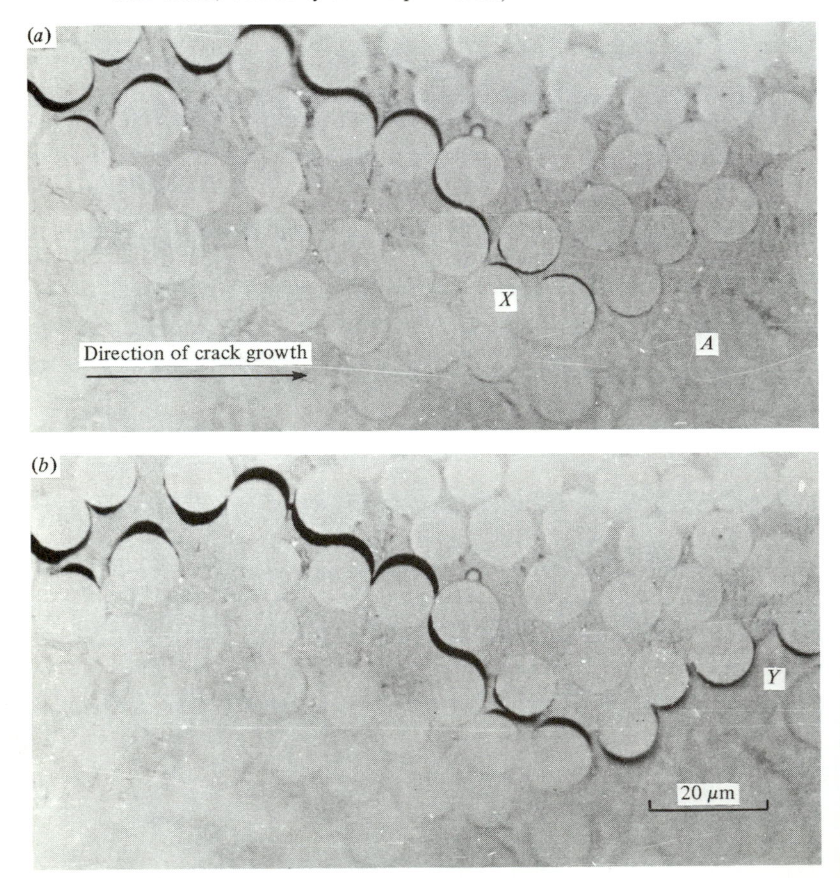

the fibres are closely spaced. The effect of resin flexibility on the transverse strength is illustrated in Fig. 7.19 for a glass fibre–polyester resin system. The stress–strain curves of the resins range from completely brittle behaviour with the strain to failure less than 0.02 to flexible behaviour with strains to failure of 0.05. There is a corresponding reduction in the modulus and strength of the resin. The transverse strains to failure of laminae with $V_f \approx 0.5$ for these resins varies between 0.0018 and 0.0026. There is a small amount of non-linear behaviour when the matrix is very flexible but the overall response appears to be brittle. For $V_f \approx 0.5$ the strain magnification from equation (5.23) is approximately 5. This is much less than that required to explain the difference between the strain to failure of the resin and the laminae. However, the fibres are invariably not uniformly distributed as assumed in deriving equation (5.23) and there are local regions where the fibres are more closely spaced, and in some places touching (Fig. 7.23). It follows from the strain magnification approach that the transverse strength will decrease with increasing V_f.

Microscopic studies of the fracture of laminae confirm that transverse cracks are nucleated in regions of dense packing and also propagate preferentially through such regions. The sequence of photographs in Fig. 7.23 illustrates these and other features of transverse cracking. The tests were done under conditions where catastrophic fracture from the first crack nucleus was avoided and it was possible to produce both unstable and stable crack growth. The main crack in Fig. 7.23 follows the fibre–matrix interface in regions of dense packing. Cracks can nucleate ahead of the main crack by debonding or by resin fracture very close to the fibre–matrix interface in regions of maximum radial tensile stress or strain, (e.g. *X* in Fig. 7.23).

In unstable crack growth, the debonding cracks propagate across the resin bridges between the fibres in a brittle mode whereas during stable crack growth many debonding cracks develop ahead of the main crack. Plastic deformation occurs in the resin bridges (e.g. *Y* in Fig. 7.23) and eventually the debonding cracks are joined by shear or tearing cracks. The main crack has been deflected around the resin rich area at *A*. These effects are strongly dependent on crack speed which affects the local strain rate in the resin at the crack tip. Under some conditions large amounts of plastic flow occur in the resin as shown in Fig. 7.24.

The fracture surface morphology is consistent with these observations on crack nucleation and growth. A typical example of a transverse tensile fracture in a carbon fibre–epoxy resin system is shown in Fig. 7.25. The crack has followed the fibre leaving clean fibre surfaces and has also propagated through the resin matrix. The river line markings in the resin

Fig. 7.24. Propagation of a transverse crack in a glass fibre–polyester resin lamina. Crack has intersected a resin rich region and extensive crack blunting and resin flow has occurred. (From M. L. C. Jones, PhD thesis, University of Liverpool 1981.)

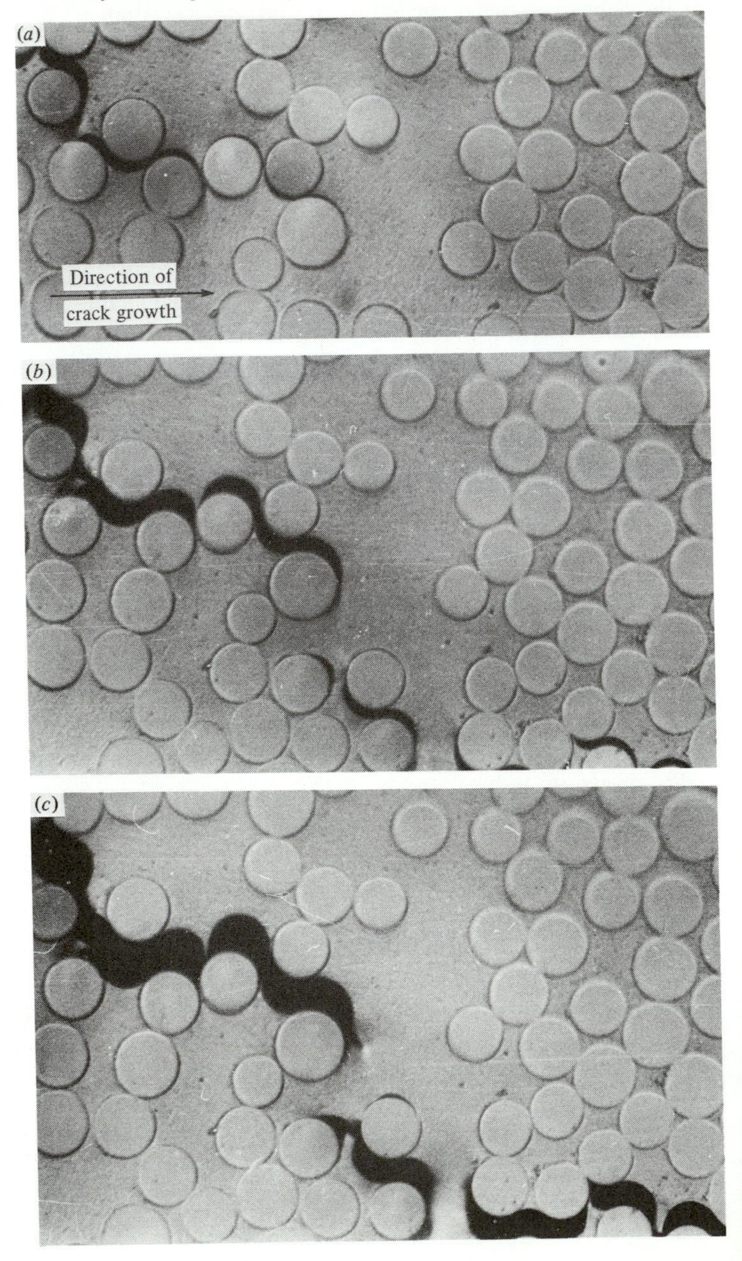

are characteristic of brittle fracture and can be used to determine the direction of crack growth. The details of the fracture surface morphology will depend on the fibre–resin system and the specific mechanisms of crack nucleation and propagation. Hence, fractography can be an important diagnostic tool in fracture analysis.

Other sources of low transverse tensile strength include debonding at the interface before cohesive failure of the resin and cohesive failure of the fibre. The latter may be particularly relevant to carbon fibres which have layer structures oriented parallel to the surface. In both cases sharp cracks may be formed which can lead to catastrophic failure. In addition, shrinkage during curing and differential thermal contraction

Fig. 7.25. Scanning electron micrograph of a transverse tensile fracture in a unidirectional carbon fibre–epoxy resin lamina (From Chamis 1979.)

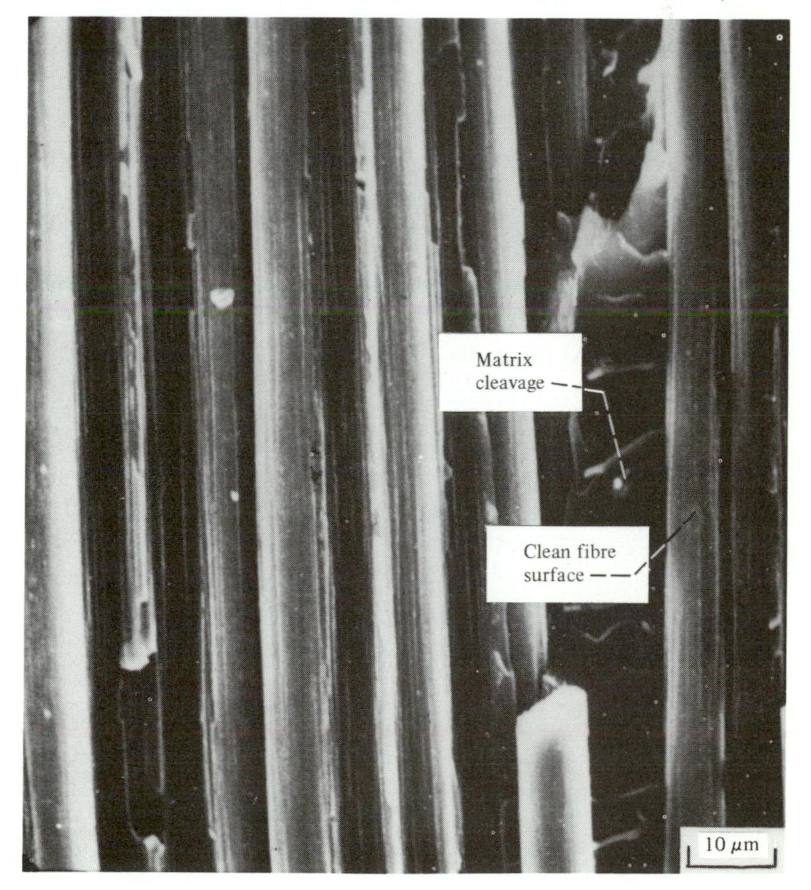

can lead to transverse cracking in laminate structures even in the absence of an external load.

Because of the importance of achieving some transverse strength considerable effort has been devoted to modifying the matrix structure so as to minimise the stress concentration effects. Two possibilities are of interest: (i) introduction of a very fine dispersion of rubber particles into the brittle resin; these are known to increase the fracture toughness of rigid resins without too big a reduction in strength and stiffness; this effect is likely to be of most value in materials with low V_f, (ii) use of an intermediate layer at the fibre–matrix interface which may result in a change in the stress pattern and a reduction in the stress magnification. The value of both these approaches in commercially important composites with high V_f where the fibres are almost touching has yet to be demonstrated. It will be noted from Fig. 4.4 and Table 4.1 that at high V_f even a thin interlayer takes up a large proportion of the total resin volume.

7.5 Longitudinal compressive strength

There is no really satisfactory account of compressive failure. A number of failure modes have been proposed and there is some experimental evidence available. Like the transverse tensile and shear strength of laminae, the longitudinal compressive strength is dependent on many factors including the fibre and resin properties, the interface bond strength and void content. The failure mode is particularly sensitive to the volume fraction of fibres and the resin properties.

The usual starting point for prediction of longitudinal compressive strength properties in terms of the properties of the fibre and resin is work by Rosen (1965) who envisaged that failure under these loading conditions would be associated with the buckling of fibres as in column buckling of struts. The buckling is restricted by the surrounding matrix so that the buckling stress and hence the longitudinal compressive strength will depend on the elastic properties of the matrix. Rosen modelled the compressive failure of a unidirectional lamina by a two-dimensional model as illustrated in Fig. 7.26. At low V_f the extensional or out-of-phase mode of buckling is predicted and the compressive strength is given by

$$\sigma^*_{\parallel C} = 2V_f[V_f E_m E_f/3(1-V_f)]^{\frac{1}{2}} \qquad (7.26)$$

The restriction to low V_f means that this mode is not relevant to commercially important composite materials. At high V_f the shear or in-phase mode of buckling is predicted giving

$$\sigma^*_{\parallel C} = G_m/(1-V_f) \qquad (7.27)$$

This indicates that $\sigma^*_{\parallel C}$ increases with V_f but is dominated by the shear modulus of the matrix. The derivation of equations (7.26) and (7.27) is given in Holister & Thomas (1966). A comparison of the values of $\sigma^*_{\parallel C}$ quoted in Table 7.1 with those predicted by equation (7.27) is given in Table 7.2. The agreement is poor, the predicted values being much greater than the experimental results. This suggests that either the microbuckling model is completely wrong or that the underlying assumptions have been seriously over-simplified.

It seems likely that the basic idea of co-operative buckling shown in Fig. 7.26*b* is valid for some fibre–matrix systems but requires modification to take account of material effects which invalidate some of the assumptions used in predicting equation (7.27).

Table 7.2. *Comparison of experimental and predicted values of longitudinal compressive strength of unidirectional laminae,* $V_f \approx 0.50$

Material	$\sigma^*_{\parallel C}$ experimental ($MN\,m^{-2}$)	$\sigma^*_{\parallel C}$ predicted (eq. (7.26)) ($MN\,m^{-2}$)	$\sigma^*_{\parallel C}$ predicted (eq. (7.27)) ($MN\,m^{-2}$)
Glass–polyester	600–1000	8700	2200
Type I carbon–epoxy	700–900	22800	2900
Kevlar 49–epoxy	240–290	13200	2900

Fig. 7.26. Schematic representation of (*a*) out-of-phase and (*b*) in-phase modes of buckling in unidirectional laminae in longitudinal compression.

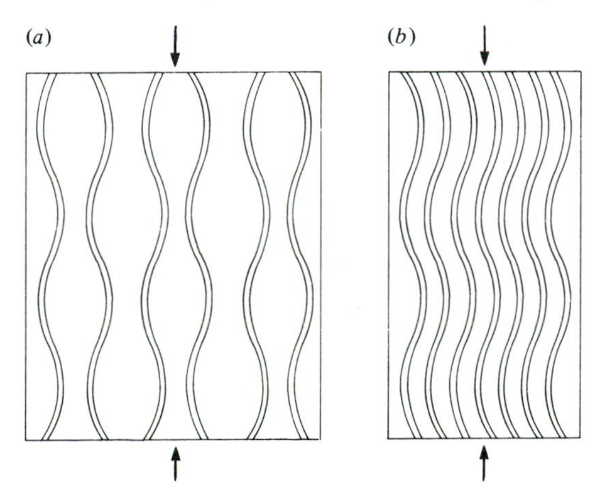

In addition, there is evidence for a completely different mode of failure in some fibre resin systems as, for example, in carbon fibre–epoxy resin materials where failure is dominated by fibre strength. At the end of this section an example is given of the transition from one mode of failure to another.

Any factor which leads to a reduction in the support that the matrix and the surrounding fibres give to the fibre to prevent microbuckling will lead to a reduction in $\sigma^*_{\parallel C}$ according to equation (7.27). In real systems this support can be reduced by the following effects.

(*a*) Fibre bunching which results in local resin rich regions. The resistance to buckling in these regions is less than that predicted for a uniform arrangement with the same V_f. Compressive failure may initiate in these regions.

(*b*) Presence of voids which have an even more pronounced effect than resin rich regions and, depending on their size, can lead to a significant reduction in compressive strength.

(*c*) Poor alignment of fibres which results in some fibres being preferentially oriented for easy buckling.

(*d*) Fibre debonding which may occur as a result of differences between the Poisson ratios of the matrix and fibre. Thus for example, when a load is applied parallel to the fibres, tensile stresses are generated at some positions around the circumference of the fibre depending on the fibre arrangement. Shrinkage stresses due to curing and differential thermal contraction add to this effect. Debonded fibres buckle more readily than well-bonded fibres.

(*e*) Viscoelastic deformation of the matrix which results in an effective matrix shear modulus less than the instantaneous modulus. All these effects, singly or in combination, mean that the predictions of equation (7.27) over-estimate $\sigma^*_{\parallel C}$. It follows also that $\sigma^*_{\parallel C}$ is dependent on manufacturing conditions, since these have a strong effect on many of these variables.

An additional factor which applies to carbon and Kevlar 49 is the non-isotropic elastic properties of these fibres. The derivation of equation (7.27) assumes that both the fibre and the matrix are isotropic. The results in Table 7.2 show that the discrepancy between the experimental values of $\sigma^*_{\parallel C}$ and the predictions of equation (7.27) are much greater for carbon and, in particular, Kevlar 49 fibres than for glass fibres. Both carbon and Kevlar 49 fibres have low transverse and shear moduli which means that predictions based on isotropic properties will over-estimate the buckling strength of the composite material. In addition, the low compressive strength of Kevlar 49 results in premature failure of the fibres by yielding.

The fracture surface usually has many longitudinal splits, and some splaying of the fibres occurs. The amount of splaying depends on the interface bond strength. In strongly bonded materials the fracture processes in the matrix are dominated by shear. The fibres fracture in a bending mode as illustrated in Fig. 7.27. Bending induces tensile and compressive stresses across the fibre. In glass and carbon fibres final

Fig. 7.27. (*a*) Tensile and compressive stresses in a fibre due to in-phase buckling leading to a kink zone, (*b*) two planes of fracture formed with brittle carbon fibres (*c*) unfractured kink zone formed with Kevlar 49 fibres.

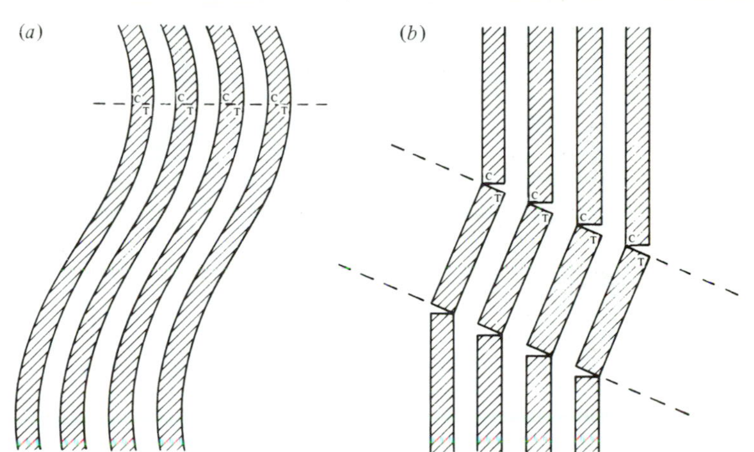

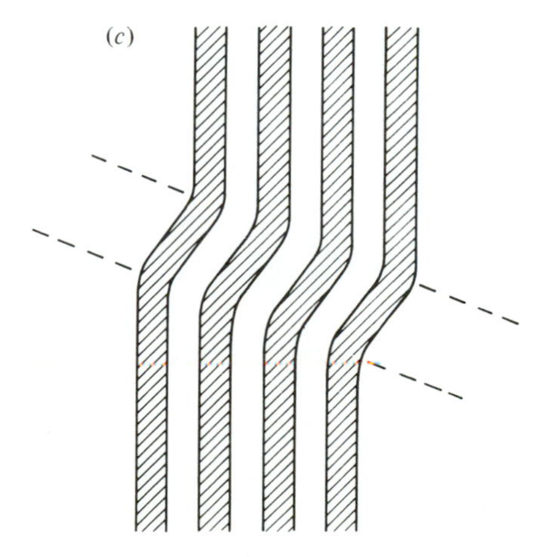

Fig. 7.28. Scanning electron micrographs of the fracture surface of a carbon fibre–epoxy resin lamina after a fibre buckling mode failure due to a longitudinal compressive stress. (*a*) Low magnification view showing smooth fracture surface. (*b*) High magnification view showing tension and compressive fracture in a single fibre. (From Ewins & Potter 1980. Crown copyright: the material is used with the permission of the Controller of Her Majesty's Stationery Office.)

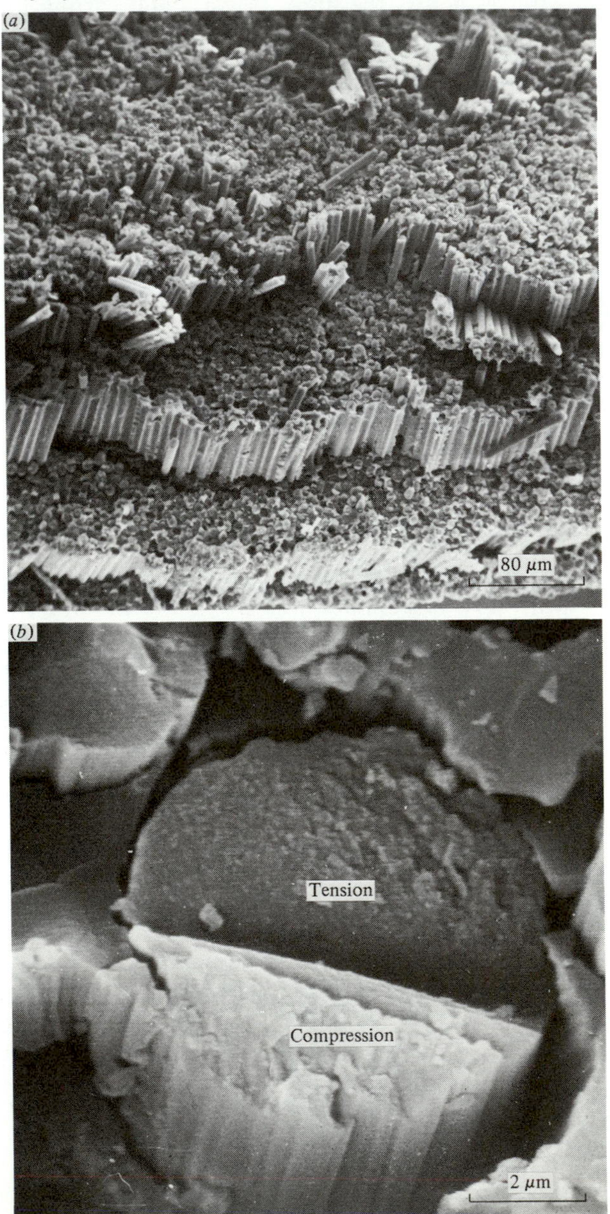

failure will occur by brittle fracture in the tensile region of each fibre followed by crushing or shear failure in the compressive region. This effect is shown very clearly in Fig. 7.28 for a carbon fibre–epoxy resin lamina. The tensile and compressive regions in the fibre are readily distinguished by the differences in fracture morphology.

Longitudinal compressive fracture can be demonstrated readily in three-point and four-point bend tests and is often associated with the development of kink zones as illustrated in Fig. 7.29. The fibres have fractured along two planes and massive fibre rotation has occurred between the planes. Similar observations have been made with Kevlar 49 fibres but little fibre fracture occurs and the kinks are due to compressive yield in the fibres.

In addition to compressive failure due to buckling, Ewins & Ham (1973) have proposed that the shear stresses generated in the laminae may be sufficient to cause a shear mode of failure as illustrated in Fig. 7.30. The shear stress due to the compressive load is given by

$$\tau = \sigma_{\parallel C} \sin\theta \cos\theta \tag{7.28}$$

which has a maximum value on a plane $\theta = 45°$ given by

$$\tau_{max} = \tfrac{1}{2}\sigma_{\parallel C} \tag{7.29}$$

Fig. 7.29. Polished section of a carbon fibre–epoxy resin pultruded specimen showing a kink band formed in compressive zone of a four-point bend test sample. (From Parry & Wronski 1980.)

If the shear strength is less than the buckling strength shear failure will occur in preference to buckling. It is possible that shear failure of individual carbon fibres can be initiated in local regions of misaligned crystallites associated with flaws introduced in manufacture. The strength will follow a rule of mixtures equation

$$\sigma^*_{\parallel C} = 2[V_f\tau^*_f + (1 - V_f)\tau^*_m] \tag{7.30}$$

where τ^*_f and τ^*_m are the shear strengths of the fibre and resin matrix respectively. Some results are shown in Fig. 7.31. The fracture surface morphology of shear fractured laminae is completely different from compressive buckling fractures. The surface is relatively smooth and the fibre ends are ill defined. A large amount of debris, consisting of particles of shattered matrix and fibres, forms on the fracture surface.

In carbon fibre–epoxy resin systems there is a transition from shear mode failure to fibre buckling failure which is dependent on the shear modulus of the matrix and the shear strength of the fibres. This can be demonstrated by testing over a range of temperatures as illustrated in Fig. 7.32. The modulus of the matrix decreases rapidly with increasing temperature and this means, following equation (7.27), that the fibre buckling mode is favoured, at high temperatures. A corresponding change will occur when the matrix modulus is modified in other ways. Thus, for example, the up-take of water leads to a significant reduction

Fig. 7.30. Schematic view of shear failure due to longitudinal compressive stresses.

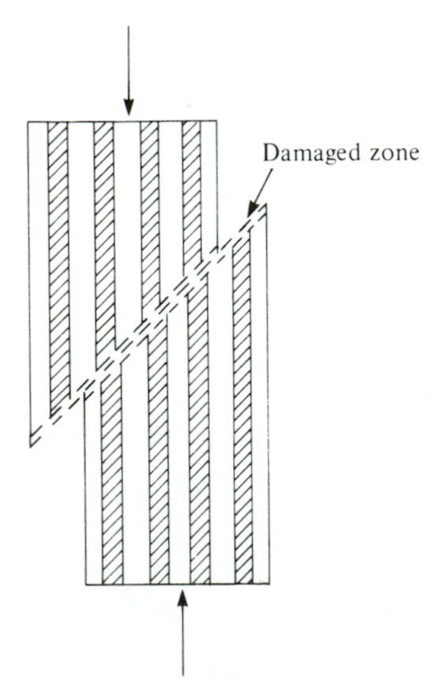

Fig. 7.31. Effect of V_f on longitudinal compressive strength of carbon fibre–epoxy resin laminae at 20 °C (From Ewins & Ham 1973. Crown copyright: the material is used with the permission of the Controller of Her Majesty's Stationery Office.)

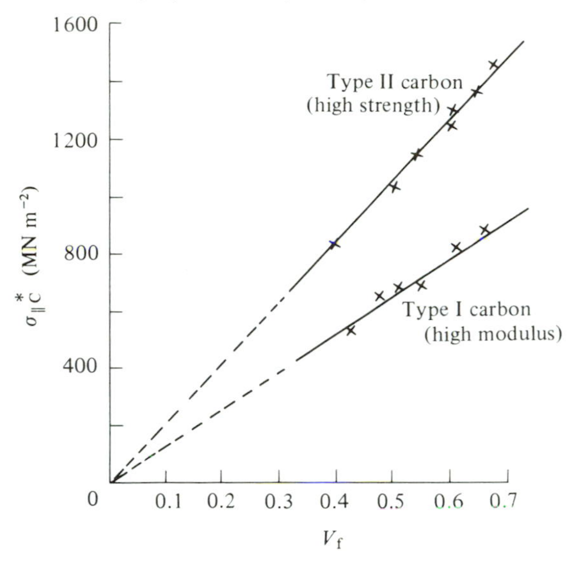

Fig. 7.32. Variation of longitudinal compressive strength of carbon fibre–epoxy resin laminae with temperature showing transition from shear mode to buckling mode failure. (After Ewins & Potter 1980. Crown copyright: the material is used with the permission of the Controller of Her Majesty's Stationery Office.)

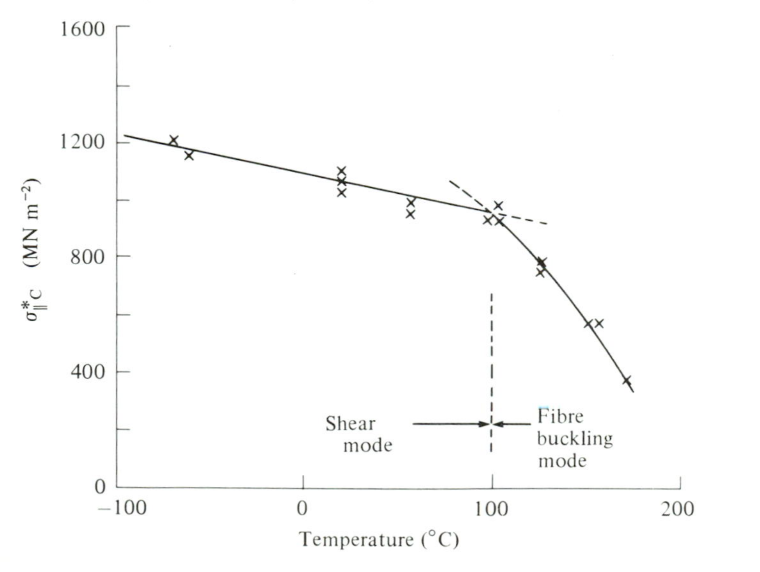

in resin modulus and an increasing tendency to fibre buckling failure. A similar effect results from an increase in fibre strength.

7.6 Transverse compressive strength

Although the maximum shear stress operates on all surfaces at 45° to the compressive axis (equation (7.29)), reference to Fig. 7.33 shows that the arrangement of fibres in the maximum shear stress surface is dependent on fibre orientation. Shear failure occurs preferentially on planes containing the fibre direction, e.g. *ABCD* in Fig. 7.33. The shear strength on this plane will be higher than the shear strength parallel to the fibres, described in the next section. This is due to the difficulty of sliding the shear surfaces over each other particularly when there is a compressive stress acting on the surface.

Shear on plane *BEDF* in Fig. 7.33 can be induced by compression by testing in such a way that displacements parallel to *FD* are prevented. In this case shear can occur only by fibre fracture as illustrated in Fig. 7.30 for longitudinal compressive failure. For carbon fibre–epoxy resin materials, it is found that the shear strengths in the two modes are similar.

Fig. 7.33. Transverse compression of a unidirectional lamina. Shear on *ABCD* can occur without fibre fracture; shear on *BEDF* must involve fibre fracture.

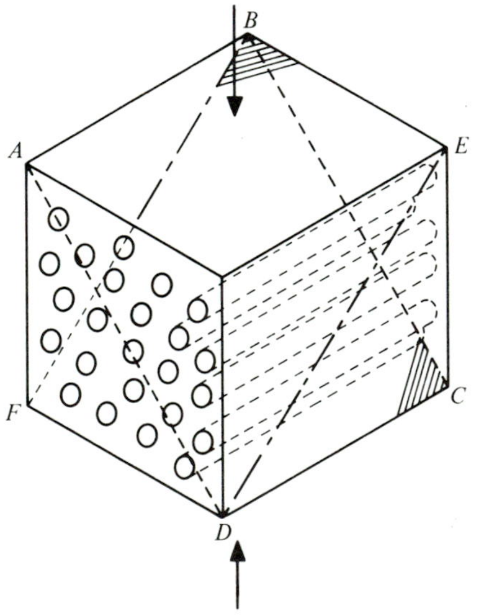

7.7 In-plane shear strength

The in-plane shear strength, or intralaminar shear strength depends on the direction of the shear displacements. The two extremes correspond to shear stresses τ_{12} and τ_{21} applied parallel to the fibres and perpendicular to the fibres; see Fig. 5.1*a*. Shear failure associated with τ_{21} resembles failure on plane *BEDF* illustrated in Fig. 7.33 except that there is no compressive stress acting.

The most relevant shear strength is τ_{12}^* or $\tau_{\#}^*$ since it is likely to be the lowest in-plane shear strength. The strength is dominated by matrix properties because crack propagation can occur entirely by shear of the matrix without disturbing or fracturing the fibres. Thus $\tau_{\#}^*$ will be dependent on the viscoelastic properties of the resin in a similar way to the transverse tensile properties described in Section 7.4. A typical shear stress–shear strain curve is shown in Fig. 7.34 for a glass fibre–polyester resin lamina. The large deviation from linearity is due to the properties of the resin.

For a given resin matrix, the shear strength depends on the stress concentration effects associated with the presence of fibres and voids and on the strength of the interfacial bond. The effect of V_f on the stress concentration factor is illustrated in Fig. 7.35 for a system with $G_f/G_m \approx 20$. At low V_f the stress concentration factor is relatively

Fig. 7.34. Shear stress–shear strain curve of a unidirectional lamina of glass fibre–polyester resin, $V_f = 0.40$. (From M. J. Legg, PhD thesis, University of Liverpool 1980.)

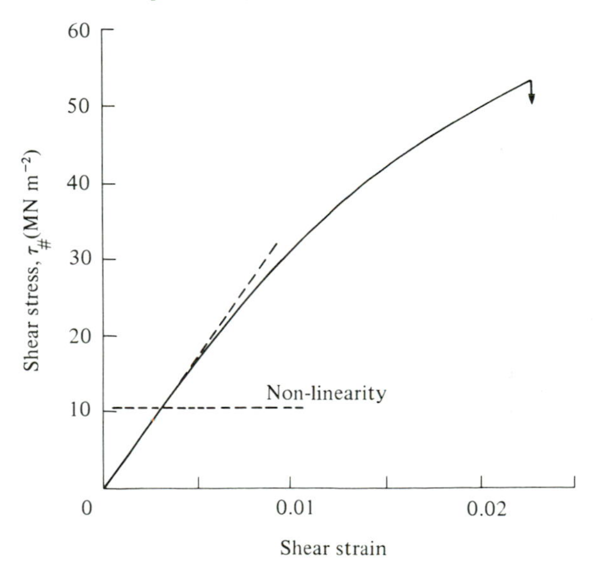

insensitive to V_f but it rises rapidly when V_f is greater than 0.60. For laminae with brittle resins, the stress concentration effect will lead to values of $\tau_{\#}^*$ lower than the shear strength of the pure resin. The effect is more pronounced when fibre bunching occurs. The strength of laminae with more flexible resins is approximately the same as the pure resin because any stress concentrations are relaxed by local deformation processes. The same arguments apply to the effect of voids and weakly bonded interfaces.

An edge view of an intralaminar shear crack is shown in Fig. 7.36. Failure has occurred by shear failure of the resin and fracture close to or at the fibre matrix interface. Shear zones have grown ahead of the crack and some local interface debonding has occurred. The fracture morphology of a shear fracture is shown in Fig. 7.37 and may be compared with the morphology of a transverse crack for the same fibre–matrix system (see Fig. 7.25). The most significant difference is the mode of failure of the resin.

7.8 Orientation dependence of strength and failure criteria

In the preceding sections the strengths of laminae in the principal material directions have been considered. In most practical applications the laminae will be subject to biaxial and triaxial loads. It is necessary, therefore, to have some method of predicting failure in such conditions. In this section the prediction of the failure strength is

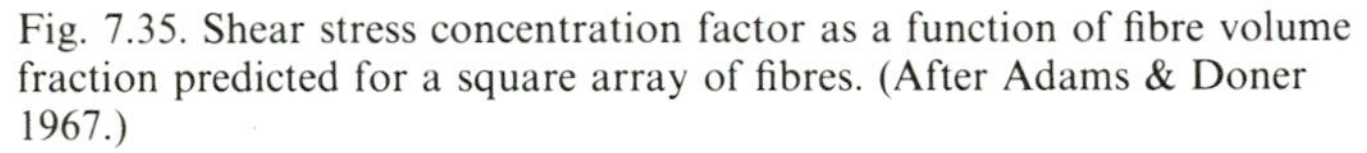

Fig. 7.35. Shear stress concentration factor as a function of fibre volume fraction predicted for a square array of fibres. (After Adams & Doner 1967.)

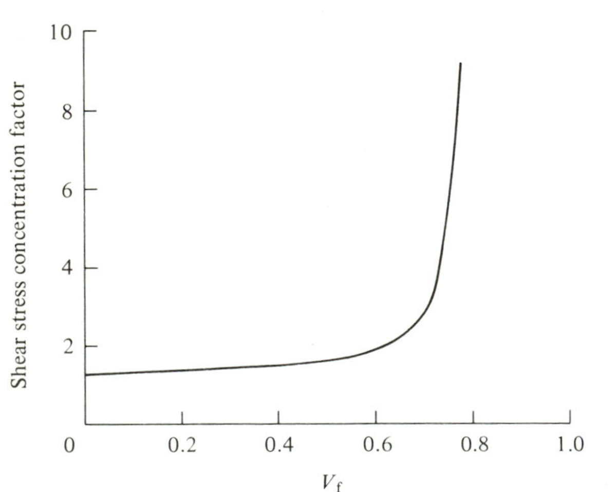

considered with reference to two simple test conditions, i.e. (i) off-axis uniaxial tensile tests, and (ii) combined uniaxial tensile and shear tests.

Following the previous discussion three simple failure modes can be envisaged in a lamina tested in uniaxial tension at an angle θ to the fibre direction. These are transverse tensile fracture, longitudinal tensile fracture and intralaminar shear fracture. The *maximum stress theory* of

Fig. 7.36. Intralaminar shear crack in a glass fibre–polyester resin lamina (From M. L. C. Jones, PhD thesis, University of Liverpool 1981.)

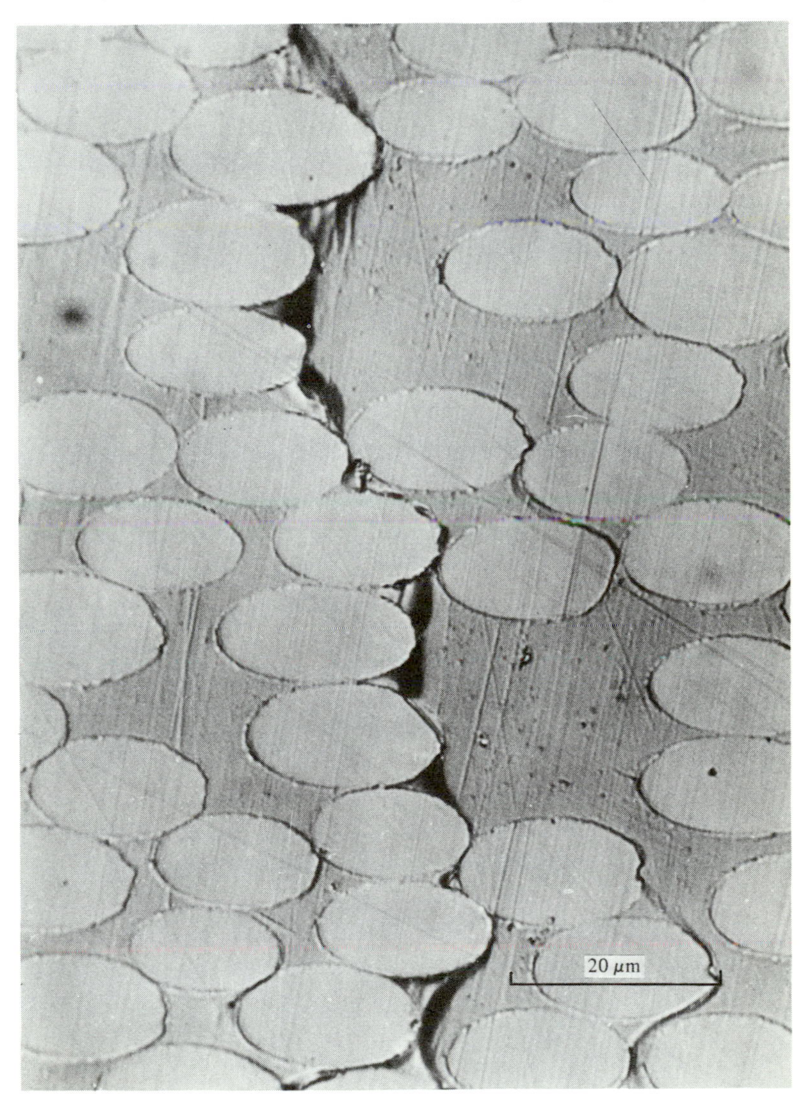

fracture assumes that fracture occurs when the stresses in the principal material directions reach a critical value. Since there are three possible failure modes there are three separate failure conditions which can be written

$$\sigma_{\parallel} = \sigma_{\parallel}^{*}, \quad \sigma_{\perp} = \sigma_{\perp}^{*}, \quad \tau_{\#} = \tau_{\#}^{*} \tag{7.31}$$

The values of $\sigma_{\parallel}$, $\sigma_{\perp}$ and $\tau_{\#}$ for a given applied stress σ_{θ} can be obtained by transformation and are given by

$$\left.\begin{aligned} \sigma_{\parallel} &= \sigma_{\theta} \cos^2 \theta \\ \sigma_{\perp} &= \sigma_{\theta} \sin^2 \theta \\ \tau_{\#} &= -\sigma_{\theta} \sin \theta \cos \theta \end{aligned}\right\} \tag{7.32}$$

At $\theta = 0°$, $\sigma_{\parallel} = \sigma_{\theta}$ and failure by longitudinal fracture occurs. Similarly,

Fig. 7.37. Scanning electron micrograph of an intralaminar shear fracture in a unidirectional carbon fibre–epoxy resin lamina. (From Chamis 1979.)

at $\theta = 90°$, $\sigma_\perp = \sigma_\theta$ and transverse fracture occurs. At intermediate angles the failure strength predicted by the maximum stress theory depends on the relative values of $\sigma_\parallel^*$, $\sigma_\perp^*$ and $\tau_\#^*$ and is obtained from the smallest value of the following stresses

$$\left.\begin{aligned} \sigma_\theta &= \sigma_\parallel^*/\cos^2\theta \\ \sigma_\theta &= \sigma_\perp^*/\sin^2\theta \\ \sigma_\theta &= \tau_\#^*/\sin\theta\cos\theta \end{aligned}\right\} \tag{7.33}$$

As an example, consider the failure strength of a glass fibre–polyester resin system for values of θ between 0° and 90°. For this system $\sigma_\parallel^* = 700$ MN m^{-2}, $\sigma_\perp^* = 22$ MN m^{-2} and $\tau_\#^* = 50$ MN m^{-2} (Table 7.1). Using equation (7.33) the failure curves shown in Fig. 7.38 are obtained. Thus, the theory predicts that longitudinal tensile failure will

Fig. 7.38. Orientation dependence of fracture strength for off-axis tests on unidirectional laminae predicted by maximum stress theory. Data for $\sigma_\parallel^*$, $\sigma_\perp^*$ and $\tau_\#^*$ from tests on glass fibre–polyester resin.

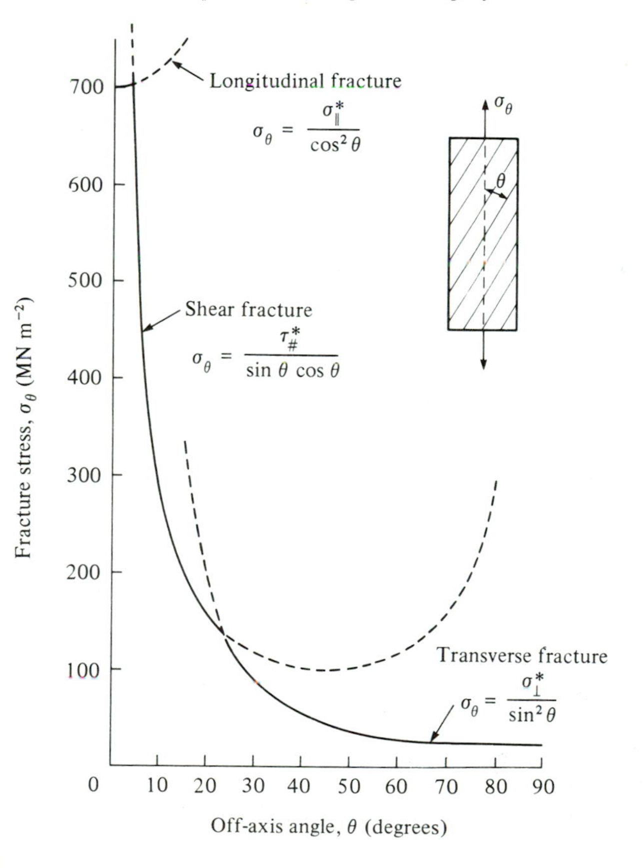

occur at $0° < \theta < 4°$, shear failure will occur at $4° < \theta < 24°$ and transverse tensile fracture will occur at $\theta > 24°$. Clearly, the angles at which the transitions from one fracture mode to another occur depend on the relative magnitudes of $\sigma_{\parallel}^*$, $\sigma_{\perp}^*$ and $\tau_{\#}^*$.

A typical set of results from off-axis tests is shown in Fig. 7.39 for a carbon fibre–epoxy resin lamina with $V_f = 0.5$. Great care is necessary in measuring the off-axis strength to avoid or at least minimise the bending forces associated with grip constraints, particularly at small values of θ. Experimentally it was found (Sinclair & Chamis, 1979) that for this material longitudinal tensile fracture occurred close to $\theta = 0°$, intralaminar shear occurred at $5° < \theta < 20°$ and transverse tensile fracture occurred at $45° < \theta < 90°$. In the range $20° < \theta < 45°$ a mixed mode fracture with intralaminar shear and transverse tensile fracture was observed. Fig. 7.39 shows that there is excellent agreement between the maximum stress theory and experimental values of the fracture strength except in the range $18° < \theta < 42°$. Outside this range it is clear that either the transverse stress or the shear stress is dominant and the maximum stress theory is perfectly adequate. However, the relatively

Fig. 7.39. Orientation dependence of fracture strength for off-axis tests on unidirectional laminae. Experimental results from tests on Type I carbon fibre–epoxy resin, $V_f \approx 0.5$. (Data from Sinclair & Chamis 1979.)

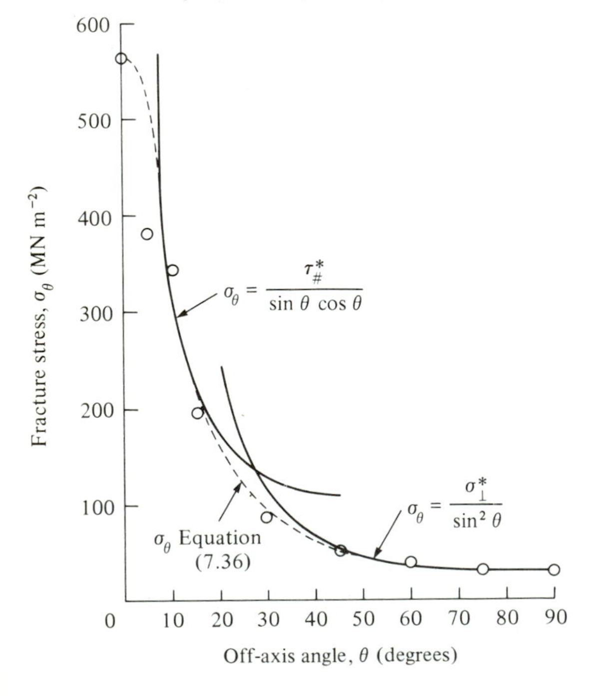

poor agreement in the intermediate range, which has been observed in many experimental investigations highlights the inadequacy of the theory. In fact, the off-axis test is not very discriminating in distinguishing between different failure theories and the combined tensile–shear stress test described later in this section illustrates this point.

In the intermediate range there is an interaction between the tensile stresses which lead to transverse tensile cracks and the shear stresses which lead to intralaminar cracks on the same fracture plane. The results in Fig. 7.39 show that this interaction results in lower fracture stresses than those predicted by the maximum stress theory which assumes no interaction; cf. equation (7.33). Many theories and hypotheses have been developed to predict the 'failure surface' under these combined loading conditions where interaction between different failure modes becomes significant. Although these have been reviewed and evaluated by many workers, much work is still required. It should be noted that the prediction of failure in simple isotropic materials in multiaxial stress tests raises many difficulties so it is not surprising that in highly anisotropic materials the problems are complex. A vast amount of experimental work is required to establish a failure criterion. This discussion is restricted to some relatively simple aspects of the subject.

One of the best-known failure criteria which takes account of the interactions is the maximum work theory, commonly referred to as the Tsai–Hill criterion. This is based on the von Mises failure criterion which was originally applied to homogeneous and isotropic bodies, then expanded and modified by Hill to anisotropic bodies and applied to composite materials by Tsai. The criterion may be expressed as

$$\left(\frac{\sigma_{\parallel}}{\sigma_{\parallel}^{*}}\right)^{2}-\left(\frac{\sigma_{\parallel}\sigma_{\perp}}{\sigma_{\parallel}^{*2}}\right)+\left(\frac{\sigma_{\perp}}{\sigma_{\perp}^{*}}\right)^{2}+\left(\frac{\tau_{\#}}{\tau_{\#}^{*}}\right)^{2}=1 \tag{7.34}$$

which describes a surface in three-dimensional space, i.e. a failure envelope. According to this criterion, no failure will occur provided that the values of $\sigma_{\parallel}$, $\sigma_{\perp}$ and $\tau_{\#}$ in combination give a value of the left-hand side of equation (7.34) less than unity, i.e. the values represent a point inside the failure envelope. For most composite materials and test conditions $\sigma_{\parallel}^{*} \gg \sigma_{\perp}$ so that the second term is negligible and equation (7.34) becomes

$$\left(\frac{\sigma_{\parallel}}{\sigma_{\parallel}^{*}}\right)^{2}+\left(\frac{\sigma_{\perp}}{\sigma_{\perp}^{*}}\right)^{2}+\left(\frac{\tau_{\#}}{\tau_{\#}^{*}}\right)^{2}=1 \tag{7.35}$$

Equations (7.34) and (7.35) only apply to orthotropic laminae under in-plane stress conditions. More complex equations are required to account for the multitude of interactions which occur in anisotropic materials under triaxial stress conditions.

Equation (7.34) can be used to predict the failure strength in off-axis tests on unidirectional laminae. Thus, re-writing the equation gives

$$\sigma_\theta = \left[\frac{\cos^4\theta}{\sigma_\parallel^{*2}} + \left(\frac{1}{\tau_\#^{*2}} - \frac{1}{\sigma_\parallel^{*2}}\right)\sin^2\theta\cos^2\theta + \frac{\sin^4\theta}{\sigma_\perp^{*2}}\right]^{-\frac{1}{2}} \tag{7.36}$$

This equation has been fitted to the experimental results for carbon fibre–epoxy resin laminae in Fig. 7.39. The agreement is much better than for the maximum stress theory. The predicted variation of strength with angle is smooth without any cusps. It will be noted that in the range $8° < \theta < 12°$ there is relatively little interaction between the longitudinal stress and the shear stress and between the transverse stress and the shear stress. Off-axis tests in this range of angles can be used to determine the intralaminar shear strength since, from equation (7.33)

$$\tau_\#^* = \sigma_\theta^* \sin\theta\cos\theta \tag{7.37}$$

As mentioned earlier the off-axis test is not very discriminating with regard to failure criteria. The combined tensile–shear stress test on hoop wound tubes illustrated in Fig. 7.40 is much more satisfactory in this respect. In practice the test is rather complicated because of the difficulty of gripping the tubes without producing additional constraints and the need to apply the tensile and shear stresses together in a fixed ratio. Some experimental results for a glass fibre–epoxy resin system are shown in Fig. 7.41. Since the fibres lie parallel to the hoop direction of the tube, at right angles to the axis of the tube, it follows that failure in pure tension ($\tau_\# = 0$) will occur at $\sigma_\perp^*$ and similarly failure in pure shear ($\sigma_\perp = 0$) will occur at $\tau_\#^*$. Since the maximum stress theory assumes that there is no interaction between the failure strengths in tension and shear the predicted strengths lie along two sides of the rectangle $\tau_\# = \tau_\#^*$ and $\sigma_\perp = \sigma_\perp^*$. Clearly, this is totally inadequate. For this test $\sigma_\parallel = 0$ and the Tsai–Hill criterion, equation (7.34), reduces to

$$\left(\frac{\sigma_\perp}{\sigma_\perp^*}\right)^2 + \left(\frac{\tau_\#}{\tau_\#^*}\right)^2 = 1 \tag{7.38}$$

which is the equation of an ellipse as shown in Fig. 7.41. The agreement with the experimental results is good although it is likely that further refinements are necessary.

Fig. 7.40. Combined transverse tension and shear test on hoop wound tube.

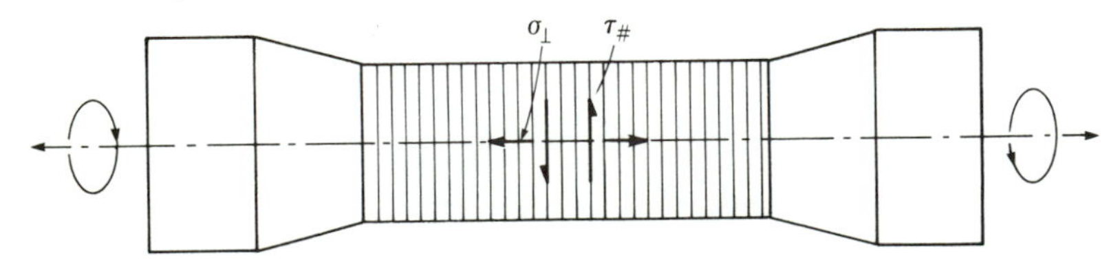

Fig. 7.41. Failure curve of unidirectional laminae of glass fibre–epoxy resin ($V_f \approx 0.65$) obtained from tests on hoop wound tubes. (From Knappe & Schneider 1973.)

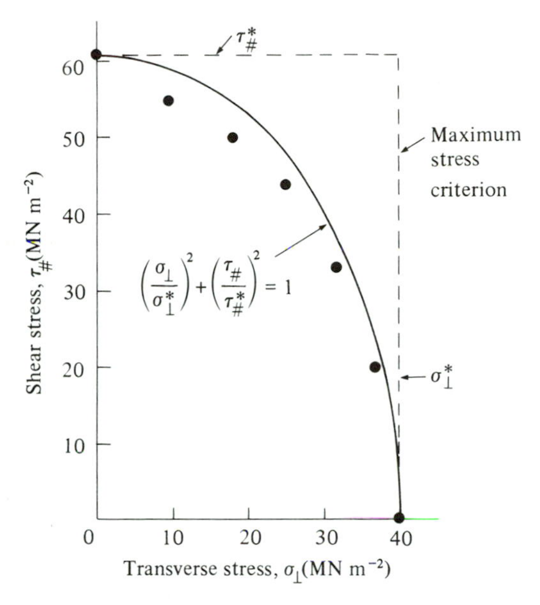

Fig. 7.42. Failure curves of unidirectional laminae of glass fibre–epoxy resin ($V_f = 0.65$) obtained from tests on hoop wound tubes. Results from short and long term tests are included. The ellipses represent the fracture criterion according to equation (7.39). (From W. Schneider, PhD thesis, University of Darmstadt 1974.)

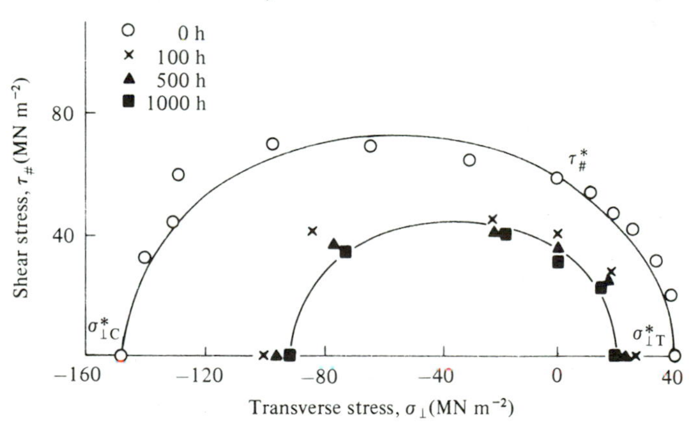

In considering failure criteria the difference between failure strengths in uniaxial tension and compression must be taken into account. This can readily be demonstrated in both the off-axis and the combined tensile–shear tests described above. One example will suffice. Fig. 7.42 shows the experimental results for combined axial compressive-shear stress tests on hoop wound tubes. The failure equation used for these results is

$$-\frac{\sigma_{\perp}^{2}}{\sigma_{\perp C}^{*}\sigma_{\perp T}^{*}}+\sigma_{\perp}\left(\frac{\sigma_{\perp C}^{*}+\sigma_{\perp T}^{*}}{\sigma_{\perp T}^{*}\sigma_{\perp C}^{*}}\right)+\left(\frac{\tau_{\#}}{\tau_{\#}^{*}}\right)=1 \tag{7.39}$$

where $\sigma_{\perp T}^{*}$ and $\sigma_{\perp C}^{*}$ are the failure strengths in transverse tension and transverse compression respectively. (The sign convention used is that stresses are written positively in the algebra and assume negative numerical values in compression.)

References and further reading

Adams, D. F. & Doner, D.R. (1967) Longitudinal shear loading of a unidirectional composite. *J. Comp. Mater.* **1**, 4–17.

Argon, A. S. (1972) Fracture of composites. *Treatise on Materials Science and Technology*, vol. 1. Academic Press, New York.

Aveston, J. & Kelly, A. (1973) Theory of multiple fracture of fibrous composites. *J. Mater. Sci.* **8**, 352–62.

Broutman, L. J. (1970) Mechanical requirements of the fiber–matrix interface. *Proceedings of the 25th SPI/RP Annual Technology Conference*, Paper 13-D. Society of the Plastics Industry, New York.

Coleman, B. D. (1958) On the strength of classical fibres and fibre bundles. *J. Mech. Phys. Solids* **7**, 60–70.

Cook, J. & Gordon, J. E. (1964) A mechanism for the control of crack propagation in all-brittle systems. *Proc. R. Soc. Lond.* A **282**, 508–20.

Corton, H. T. (1967) Micromechanics and fracture behaviour of composites. *Modern Composite Materials*, ed. L. J. Broutman & R. H. Krock, pp. 27–105. Addison-Wesley, Reading, Mass.

Ewins, P. D. & Ham, A. C. (1973) The nature of compressive failure in unidirectional carbon fibre reinforced plastics. Royal Aircraft Establishment Technical Report 73057.

Ewins, P. D. & Potter, R. T. (1980) Some observations on the nature of fibre reinforced plastics and the implications for structural design. *Phil. Trans. R. Soc. Lond.* A **294**, 507–17.

Greenwood, J. H. (1977) German work on GRP design *Composites* **8**, 175–84.

Greszczuk, L. B. (1974) Consideration of failure modes in the design of composite materials. *AGARD Conference Proceedings 163*, *Failure modes of composite materials with organic matrices and their consequences on design*, paper 12.

Hill, R. (1950) *The Mathematical Theory of Plasticity*. Oxford University Press, London.

Holister, G. S. & Thomas, C. (1966) *Fibre Reinforced Materials*. Elsevier, London.

Kelly, A. (1970) Interface effects and the work of fracture of a fibrous composite. *Proc. R. Soc. Lond.* A **319**, 95–116.

Knappe, W. & Schneider, W. (1973) The role of failure criteria in the fracture analysis of fibre-matrix composites. *Deformation and Fracture of High Polymers*, ed. H. H. Kausch, J. A. Hassell & R. I. Jaffee, pp. 543–56. Plenum Press, New York.

Outwater, J. O. & Murphy, M. C. (1969) Fracture energy of unidirectional laminates. *Proceedings of the 24th SPI/RP Conference*, paper 11-B. Society of the Plastics Industry, New York.

Parry, T. V. & Wronski, A. S. (1981) Kinking and tensile compressive and interlaminar shear failure mechanisms in CFRP beams tested in flexure. *J. Mater. Sci.* **16**, 439–50.

Puck, A. & Schneider, W. (1969) On failure mechanisms and failure criteria of filament-wound glass-fibre/resin composites. *Plast. Poly.* **37**, 33–43.

Rosen, B. W. (1965) Mechanics of composite strengthening. *Fibre Composite Materials*, ch. 3. ASM, Metals Park, Ohio.

Scop, P. M. & Argon, A. S. (1969) Statistical theory of strength of laminated composites. *J. Comp. Mater*, **3**, 30–47.

Sinclair, J. H. & Chamis, C. C. (1979) Fracture modes in off-axis fiber composites. *Proceedings of the 34th SPI/RP Annual Technology Conference*, paper 22-A. Society of the Plastics Industry, New York.

Tirosh, J., Katz, E., Lifschuetz, G. & Tetelman, A. S. (1979) The role of fibrous reinforcements well bonded and partially bonded on the transverse strength of composite materials. *Engng Fract. Mech.* **12**, 267–77.

Tsai, S. W. (1974) Structural behaviour of composite materials. NASA Contract Report CR-71.

Zweben, C. & Rosen, B. W. (1970) A statistical theory of material strength with application to composite materials. *J. Mech. Phys. Solids* **18**, 189–206.

8 Strength of laminates

8.1 Introduction

When a laminate, consisting of a stack of laminae bonded together, is subjected to a load the response depends on the properties of the individual laminae and the way they interact with each other. The elastic response has been described in Chapter 6.

The failure of laminates can be related to the strength of the laminae. The primary difficulty is defining what is meant by strength since the complete failure of the laminate is usually preceded by fracture of individual laminae. Thus, the laminate may continue to support an increasing load even though insipient failure has occurred. In practice the relevant 'strength' depends on the particular application of the composite material.

The method of approach for predicting the final failure strength of a laminate is as follows.

(i) For a given laminate construction and applied loading conditions, the stresses in the individual laminae are calculated (see Section 6.4) at progressively increasing loads.

(ii) The laminae stresses are compared with the predicted failure stresses using one of the failure criteria described in Chapter 7.

(iii) When the applied load is sufficiently large for the failure criterion to be satisfied for one of the laminae, it is assumed that failure of this lamina and all other laminae oriented in the same way occurs, and that the load supported by these laminae is transferred to the other, differently oriented, laminae. In most cases the redistribution of load is such that the residual unbroken laminae can continue to support the load.

(iv) The stresses on the remaining laminae are recalculated for further increases in load until the failure criteria is satisfied for other laminae. At this stage final failure of the laminate may occur.

Depending on the number of differently oriented laminae this analysis may have to be repeated. For a laminate with only two different orientations, final separation and failure occurs when both sets of laminae have fractured provided the applied load is sufficient to overcome the interlaminar bonding forces. This introduces an additional failure condition which is mentioned briefly at the end of this chapter. The interlaminar bond also influences the load transfer between fractured laminae so that the assumptions in (iii) above are not strictly correct. It should be noted also that the simple analysis referred to in (i) above for calculating the stresses in the laminae assumes that the stress–strain properties of the laminae are linear to failure. This is reasonable for the longitudinal properties but the transverse and shear properties (see e.g. Fig. 7.19 and 7.34) may show significant non-linearity due to flow of the resin. Hence the stresses cannot be predicted directly from the strain and more general equations are required. This is beyond the scope of this book but there is ample reference to the subject in the published literature; see Amijuma & Adachi (1979).

In this chapter the use of the approach outlined above to explain the stress–strain and failure response of some simple laminate constructions under simple loading conditions is described. In particular, reference is made to cross-ply and angle-ply laminates. As mentioned previously the experimental difficulties associated with obtaining meaningful test data for composite materials are often acute. This is highlighted in this chapter by an account of the failure of angle-ply laminates in uniaxial tension where interlaminar shear plays a major part in the failure processes, particularly for finite width specimens. The effect is strongly dependent on ply angle and width of the test specimens. Since failure of individual laminae occurs before catastrophic failure of the laminate, methods of detecting failure need to be established. In this regard reference is made to the use of microscopic studies and acoustic emission.

8.2 Cross-ply laminates

In the three layer cross-ply laminate illustrated in Fig. 8.1 the thickness of the two outer layers is the same so that the laminate is symmetrical about its mid-plane. When the laminate is loaded in uniaxial tension there are no bending moments and the stresses in the outer laminae are identical. A typical stress–strain curve of such a laminate made up of laminae with properties $\epsilon_{\parallel}^{*} > \epsilon_{\perp}^{*}$ tested in uniaxial tension parallel to the fibres in the outer layers is shown in Fig. 8.2. The most characteristic feature of the curve is the so-called 'knee'. This is associated with the formation of well-defined transverse cracks in the inner lamina which can be detected by visual examination (Figs. 8.3 and 8.4) and acoustic emission.

A useful review of the use of acoustic emission techniques in the investigation of polymer-based composite materials has been given by Sims (1976); see also Rotem (1977). Briefly, the technique involves attaching a piezo-electric transducer to the surface of the sample and recording the stress wave emission or noise produced during deformation

Fig. 8.1. Cross-ply laminate.

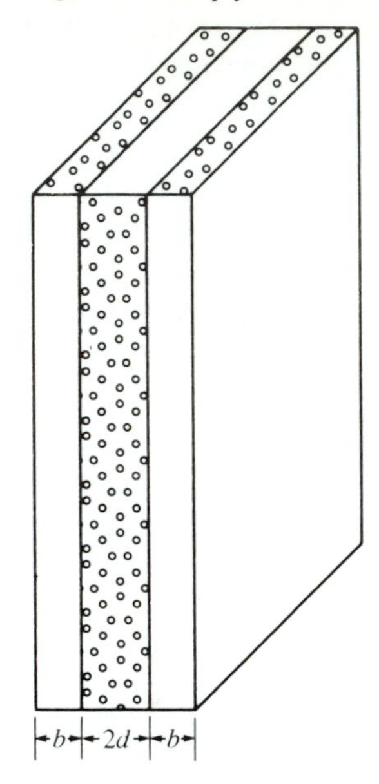

Fig. 8.2. Typical stress–strain curve and acoustic emission output of a cross-ply laminate tested in uniaxial tension.

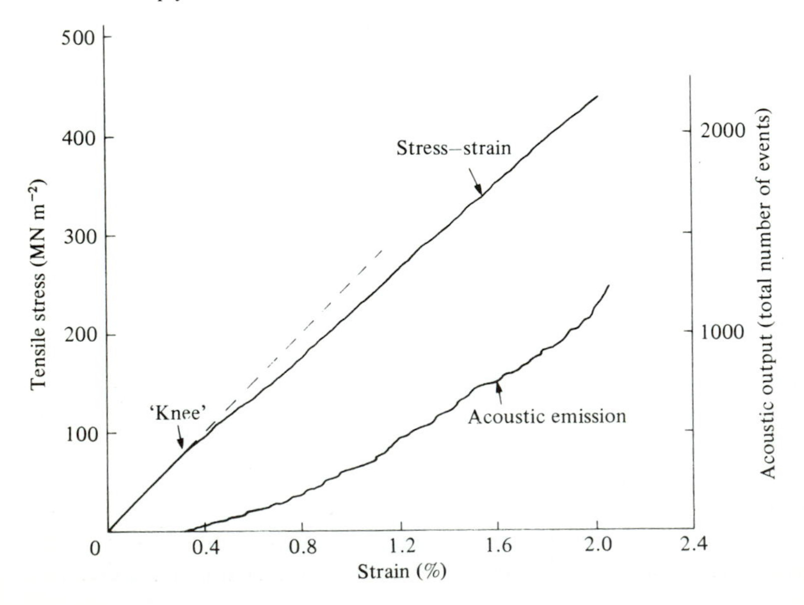

and fracture. The 'creaking' noise, which is often audible without the use of amplification equipment, is due to the generation of elastic waves by growing cracks. The acoustic emission trace shown in Fig. 8.2 is characteristic of microfracture in composite materials. There is a rapid increase in the number of noise producing events as the strain increases beyond the knee indicating that further transverse cracking is occurring. The multiple cracking process is evident in the series of photographs in Figs. 8.3 and 8.4.

The stress on the laminate at the onset of transverse cracking can be calculated, from the strength properties of the laminae and the distri-

Fig. 8.3. Transmitted light photographs of three glass fibre–polyester resin cross-ply laminates strained in tension to 1.6% showing regularly spaced transverse cracks which have formed in the centre layer. Different spacings are due to different transverse ply thicknesses (*a*) 0.75 mm, (*b*) 1.5 mm and (*c*) 2.6 mm. (From Garrett & Bailey 1977*b*.)

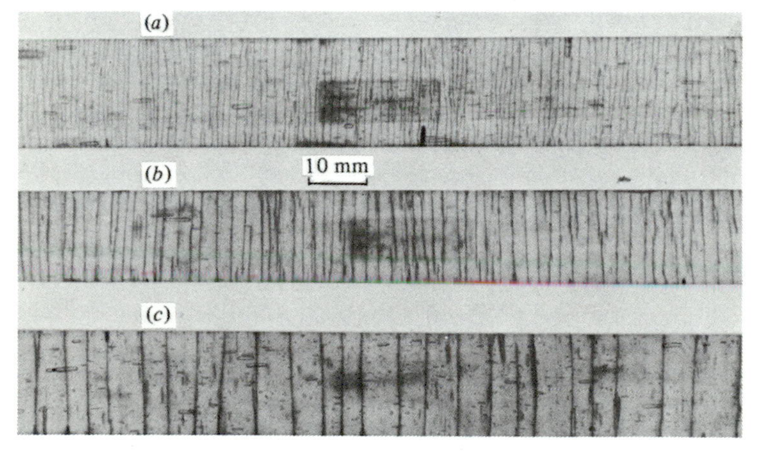

Fig. 8.4. Polished section of a symmetric 8-ply (0°, 90°) carbon fibre–epoxy resin laminate showing regularly spaced cracks in the transverse layers. (From Ewins & Potter 1980. Crown copyright: the material is used with the permission of the Controller of Her Majesty's Stationery Office.)

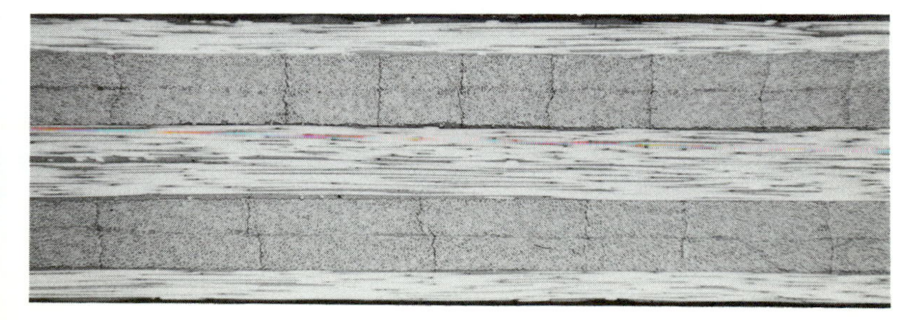

bution of the stresses in the laminate, using the failure criteria given by equation (7.35). Thus, from Section 6.4, for a laminate with the total thickness of the outer laminae the same as the thickness of the inner lamina ($b = d$ in Fig. 8.1), $V_A = V_B$, and for the inner lamina

$$^2\sigma_\perp/\sigma_{c1} = 0.180 \quad ^2\sigma_\parallel/\sigma_{c1} = -0.0373$$

when $E_1/E_2 = 10$ and $\nu_{12} = 0.25$. (The subscript c is for cross-ply and superscript 2 for 2-direction.) Since $\sigma_\parallel^* \gg \sigma_\perp^*$ it follows that the stress parallel to the fibres can be neglected and so equation (7.35) reduces to

$$(\sigma_\perp/\sigma_\perp^*)^2 = 1 \quad \text{and} \quad \sigma_\perp^* = \sigma_\perp = 0.180\ \sigma_{c1} \tag{8.1}$$

The above approach applies only when fracture of the individual laminae is not constrained by adjacent laminae. This problem has been investigated theoretically by Aveston & Kelly (1973, 1980) who showed that when the inner lamina is very thin the onset of cracking is partially or completely suppressed because insufficient elastic strain energy is released by crack nucleation and growth. It is assumed in this chapter that the laminae are sufficiently thick for this factor to be insignificant although it should be noted that the approach offers a possibility for increasing the first failure stress and strain of composite laminate structures.

Fig. 8.5. Transverse crack in a cross-ply laminate illustrating additional stress $\Delta\sigma$ on longitudinal layers.

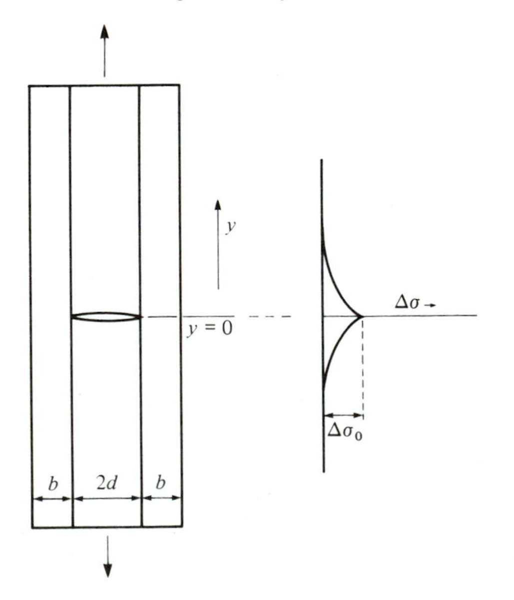

Transverse cracking of the inner lamina (Fig. 8.5) results in a redistribution of the load and an increase in the stress on the outer longitudinal laminae. Complete fracture of the laminate occurs if these stresses are greater than $\sigma_{\parallel}^*$. This depends on the thickness of the inner and outer layers and the relative values of $\sigma_{\perp}^*$ and $\sigma_{\parallel}^*$. In practical laminates complete fracture does not occur at the onset of transverse cracking because the layer thicknesses are approximately the same and $\sigma_{\parallel}^* \gg \sigma_{\perp}^*$. Thus, the final fracture strength of the laminate is determined by the fracture strength of the longitudinal layers and is given by

$$\sigma_c^* = \sigma_{\parallel}^* b/(d+b) \tag{8.2}$$

As in the case of single laminae, transverse cracking in laminates involves nucleation and propagation processes, the relative importance of which depends on the properties of the resin and the interface. Thus, for example, the well-defined cracking shown in Fig. 8.4 may be preceded by fibre debonding which produces a non-linear stress–strain response before the 'knee'.

Although the first cracks coincide with the 'knee' an increase in load is required to produce further cracks and a process of progressive or multiple cracking occurs with increasing load. The phenomena can be understood by using the multiple cracking theory developed by Aveston & Kelly (1973) which was first applied to cross-ply laminates by Garrett & Bailey (1977*a*). Briefly, the shear-lag theory (see Section 5.4) is used to predict the redistribution of stress which occurs after a transverse crack has formed. Three specific cases can be envisaged depending on the interlaminar shear properties.

(i) When there is no bonding or friction forces between the layers the occurrence of the first transverse crack results in complete unloading of the transverse lamina and no further cracking can occur. This case is of no practical interest.

(ii) When the bonding between the layers is due entirely to frictional forces there is a constant resistance to shear displacements. The stress on the longitudinal laminae immediately adjacent to the transverse crack is increased and decreases linearly away from the crack. There is a corresponding reduction in the tensile stresses in the transverse lamina. Further cracking of the transverse lamina occurs at regular intervals determined by the distance over which the stresses are reduced. Again, this case is of little practical interest for laminates except at final separation.

(iii) When the layers are elastically bonded there is again an additional stress $\Delta\sigma$ on the longitudinal layers but the rate of decay is different than for the frictional case in (ii) above (see Fig. 7.17). The additional stress has a maximum value $\Delta\sigma_0$ in the plane of the transverse crack.

According to multiple cracking theory $\Delta\sigma$ decays along the longitudinal laminae as

$$\Delta\sigma = \Delta\sigma_0 \exp(-\phi^{\frac{1}{2}} y) \tag{8.3}$$

where

$$\phi = \frac{E_c G_t}{E_{\parallel} E_{\perp}} \left(\frac{b+d}{bd^2}\right) \tag{8.4}$$

E_c is the initial composite modulus and $G_t = G_{\perp\parallel}$ is the shear modulus of the transverse layer in the y-direction (see Fig. 8.5). The corresponding load F transferred back into the transverse layer at a distance y is given by

$$F = 2bc\Delta\sigma_0 (1 - \exp(-\phi^{\frac{1}{2}} y)) \tag{8.5}$$

If the first transverse crack formed at a stress on the laminate $\sigma_c = \epsilon_{\perp}^* E_c$, which corresponds to a stress on the lamina $\sigma_{\perp}^*$, it is reasonable to assume that the next crack will occur when this stress is reached elsewhere in the transverse lamina. According to equation (8.5)

Fig. 8.6. Comparison of experimental results with theoretical curves of crack spacings as a function of applied stress for a 190 mm long glass fibre–epoxy resin cross-ply laminate with a transverse ply thickness of 1.2 mm. (From Parvizi & Bailey 1978.)

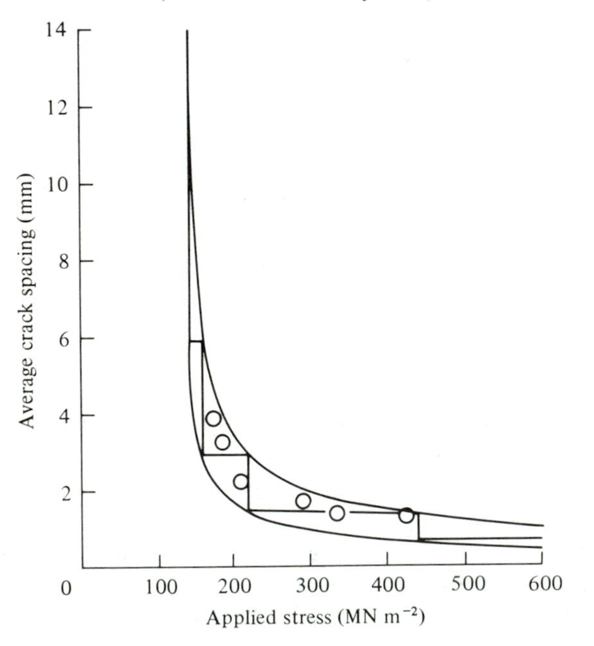

this occurs at $y = \infty$ if $\sigma_c = \epsilon_\perp^* E_c$. Thus, for another crack to form, σ_c and hence $\Delta\sigma_0$ must be increased until the stress on the transverse ply at the farthest ends of the specimen reaches $\sigma_\perp^*$. When σ_c is increased further the next series of cracks forms mid-way between the first formed cracks and the process continues until the crack spacing reaches a limiting value. An example of the change in average crack spacing with applied stress is shown in Fig. 8.6. The stepped curve represents the predicted spacing of transverse cracks when the first crack forms in the centre of 190 mm long test specimen. The next cracks form at the ends and then successive cracking occurs half-way between these cracks and so on. The upper and lower continuous curves indicate the predicted range of crack spacings for specimens of any length with an arbitrary position of the first crack. The agreement between experiment and theory is good and the photographs in Fig. 8.3 show that the crack spacing is remarkably uniform particularly when it is recognised that the strength of the transverse layer will vary along the length of the specimen because of variability in fibre packing and arrangement.

The shape of the stress–strain curve (see Fig. 8.2) of the cross-ply laminate depends on the amount of transverse cracking and the strength of the interlaminar bond. The initial slope is entirely dependent on the elastic properties of the individual laminae and their relative thicknesses; see Section 6.3. To a first approximation the simple rule of mixtures relation can be used and

$$\frac{d\sigma}{d\epsilon} = E_c = E_\parallel \frac{b}{b+d} + E_\perp \frac{d}{b+d} \tag{8.6}$$

When there is no bonding the onset of transverse cracking will result in a sharp break in the stress–strain curve. Further loading will produce a linear stress–strain curve with a slope dependent on the elastic properties of the longitudinal laminae, i.e.

$$\frac{d\sigma}{d\epsilon} = E_\parallel \frac{b}{b+d} \tag{8.7}$$

When the laminae are elastically bonded the curve will show a series of steps associated with the occurrence of transverse cracks. In practice the curve shows a relatively smooth transition (Fig. 8.2) owing to the non-uniformity of cracking, creep relaxation in the transverse laminae and the progressive, rather than instantaneous, growth of the transverse cracks. At high stress levels the curves approach the slope given by equation (8.7) because the load carrying capacity of the transverse lamina decreases as cracking increases.

In this account the constraints produced by the transverse lamina on the longitudinal laminae have been neglected. In some circumstances they lead to longitudinal cracks parallel to the fibres. This arises because the Poisson contraction of the longitudinal laminae is resisted by the high modulus fibres in the transverse lamina (see Fig. 6.15). When the tensile stress at right angles to the fibres reaches $\sigma_{\perp}^{*}$ cracks will form in the longitudinal laminae as illustrated schematically in Fig. 8.7.

8.3 Angle-ply laminates

The failure processes in angle-ply laminates are more complicated than in cross-ply laminates and they are strongly dependent on ply angle. Some aspects of the problem have been described in Section 7.8 which deals with the orientation dependence of the failure of unidirectional laminae. A particular complication which is discussed in the next section is the effect of free edges on the failure processes. Additional stresses due to interlaminar shear are present at free edges and, because of these stresses, the strengths of angle-ply laminates are dependent on the width of the test specimens.

Edge effects can be minimised by the use of very wide flat sheet

Fig. 8.7. Schematic representation of cross-ply laminate with regularly spaced cracks in transverse lamina and isolated cracks in longitudinal laminae.

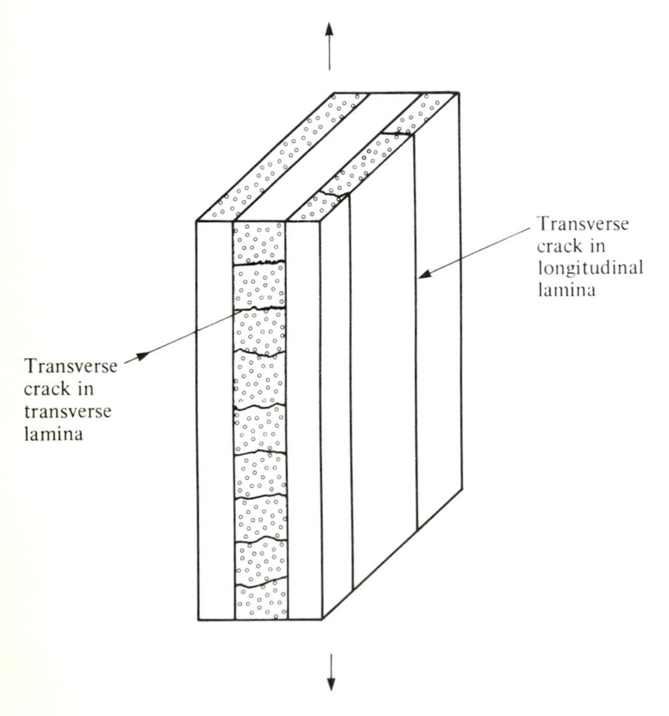

specimens, or alternatively by use of specimens in tube form. The examples in this section are based on tube tests. Two test modes are described to demonstrate the effect of uniaxial and biaxial loading conditions. These are illustrated schematically in Fig. 8.8. Uniaxial pure hoop loading (Fig. 8.8*a*) is obtained by subjecting the tube to an internal pressure without any load or restriction to the movement of the tube in the axial direction. This is achieved by containing the internal pressure with 'O' rings on which the tube is free to slide. Biaxial loading (Fig. 8.8*b*) with a hoop stress to axial stress ratio of two is obtained by pressurising the tube with one end sealed. The internal pressure acts on the surfaces of the tube to produce a hoop stress and on the sealed end to produce an axial stress. The hoop stress is given by

$$\sigma_H = P\left(\frac{2r+t}{2t}\right) \approx \frac{Pr}{t} \tag{8.8}$$

and the axial stress by

$$\sigma_A = P\left(\frac{\pi r^2}{2\pi rt}\right) = \frac{Pr}{2t} \tag{8.9}$$

therefore $\sigma_H/\sigma_A = 2$. P is the internal pressure, $2r$ is the internal diameter of the tube and t is the wall thickness of the tube. It is assumed that $r \gg t$.

Angle-ply tubes can be made by filament winding as illustrated in Fig. 8.9. The fibre tow is impregnated with resin and wound onto a mandrel. The winding angle ϕ is determined by the relative speeds of the lateral movement of the traverse and rotation of the mandrel and by the diameter of the mandrel. Accurate control of the winding process is required to ensure that the fibre bundles are laid up alongside each other.

Fig. 8.8. Schematic representation of tube tests (*a*) pure hoop loading, axial stress $\sigma_A = 0$ (*b*) hoop and end loading, hoop stress $\sigma_H = 2\sigma_A$.

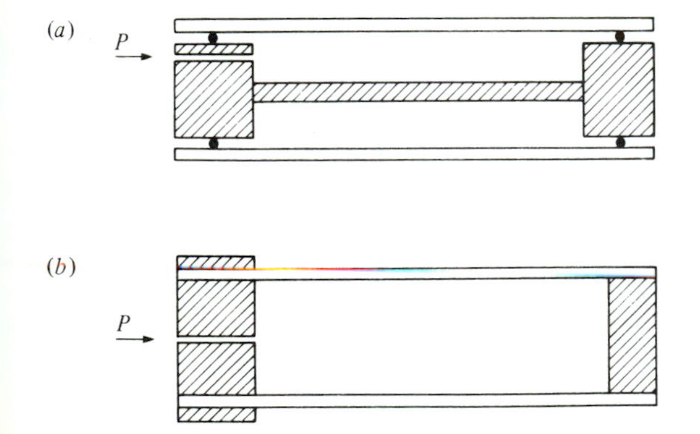

Two possible lay-up sequences are shown in Fig. 8.10. The lay-up in Fig. 8.10*a* is produced by winding a complete layer at angle θ and then overwinding with a complete layer at $-\phi$, and so on. In Fig. 8.10*b* the layers interweave as the traverse moves backwards and forwards along the mandrel.

The stresses in the individual laminae of an angle-ply laminate can be calculated by using the approach outlined in Section 6.4. Some results calculated for a glass fibre–polyester resin laminate tested in

Fig. 8.9. Schematic view of filament winding machine for making angle-ply tubes.

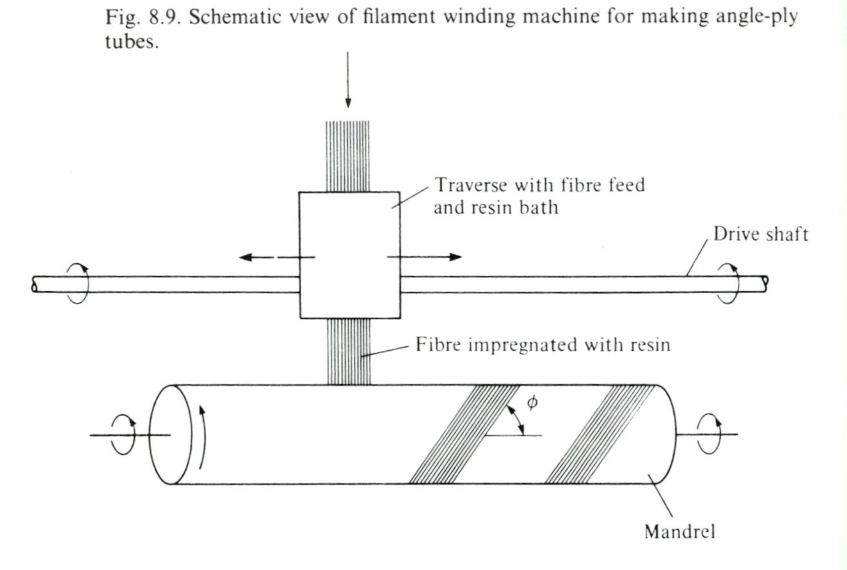

Fig. 8.10. Lay-up sequences in filament wound tubes.

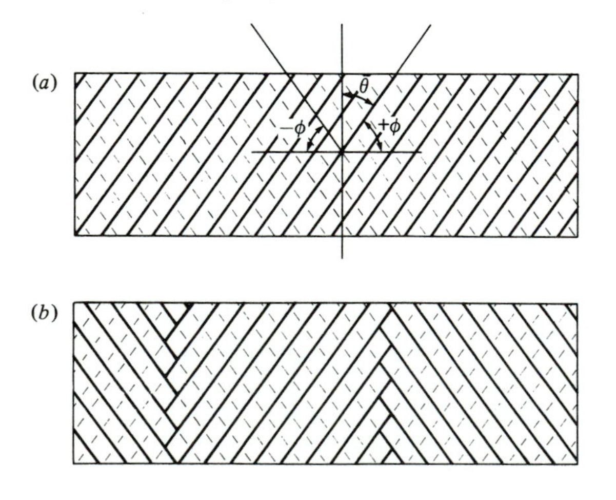

uniaxial tension are shown in Fig. 6.16. These may be compared with the corresponding stresses in a single lamina given by equation (7.32). There are some significant differences between the two sets of stresses owing to the constraints of adjacent laminae in the laminate.

The results in Fig. 6.16 have been normalised by dividing by the applied uniaxial stress, i.e. hoop stress for tubes. The most likely initial failure mode can be predicted from these curves by adopting a suitable failure criterion and using experimentally determined values of $\sigma_{\parallel}^*$, $\sigma_{\perp}^*$ and $\tau_{\#}^*$. If, for simplicity, it is assumed that the maximum stress criterion, equation (7.31), applies then the dependence of the applied stress at initial failure for an angle-ply laminate is as shown in Fig. 8.11. The results were calculated for a glass fibre–polyester resin laminate with $V_f = 0.5$ using $\sigma_{\parallel}^* = 700$ MN m^{-2}, $\sigma_{\perp}^* = 22$ MN m^{-2} and $\tau_{\#}^* = 50$ MN m^{-2}. Thus, transverse cracking will occur as the first cracking mode when $\theta = (90° - \phi)$ is greater than 43°; shear cracking will dominate where θ is less than 43°. There are, of course, the interaction effects

Fig. 8.11. Hoop stress at onset of failure predicted from Fig. 6.16 for glass fibre–polyester resin tubes ($V_f = 0.5$) tested in uniaxial tension (pure hoop stress).

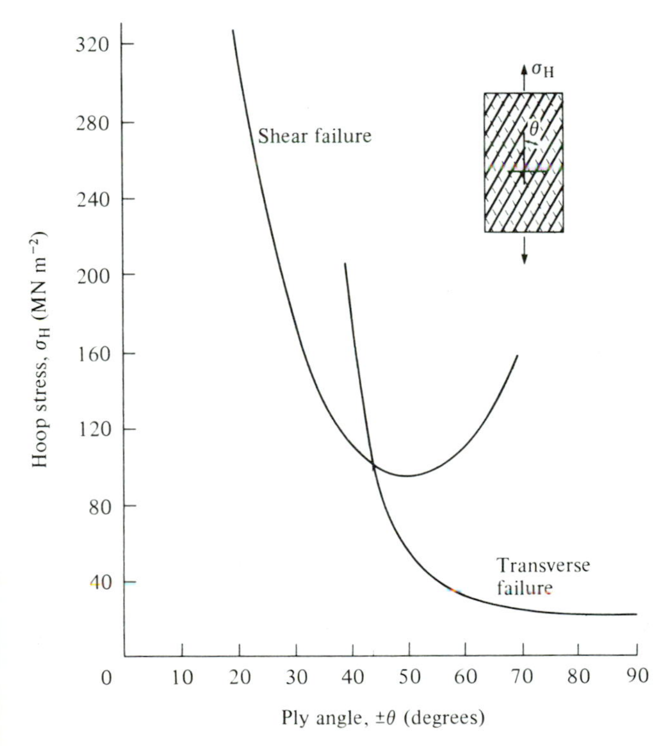

discussed in Section 7.8 which mean that a simple failure criterion is not strictly appropriate particularly when $\sigma_{\perp} \approx \tau_{\#}$.

The hoop stress–hoop strain curves of a series of angle-ply tubes tested in pure hoop loading are shown in Fig. 8.12. All the curves show an initial linear–elastic response and the slope $\sigma_H/\epsilon_H = E_{a1}$ (see Fig. 6.14). The stress at the onset of non-linearity decreases with θ. The effect can be explained by reference to Fig. 6.16 and 8.11. At large angles of θ transverse cracking is dominant and each lamina splits up into blocks in a similar way to the transverse laminae in cross-ply laminates (Fig. 8.7). As cracking proceeds the laminae become progressively uncoupled from adjacent laminae as a result of interlaminar cracking, and fibre rotation occurs by sliding of the blocks over each other. Fig. 8.12 shows that large displacements can occur. Final failure is associated with secondary effects such as fibre bending and fracture at cross-over points (Fig. 8.10*b*) and tube buckling. At lower values of θ transverse cracking is insignificant because $\sigma_{\perp}$ is very small and at $\theta < 34°$ $\sigma_{\perp}$ is negative. Non-linearity is due to shear displacements in the laminae. At low stresses the displacements are reversible indicating that the deformation

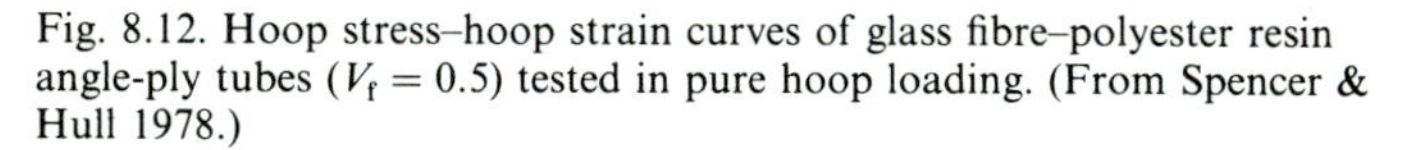

Fig. 8.12. Hoop stress–hoop strain curves of glass fibre–polyester resin angle-ply tubes ($V_f = 0.5$) tested in pure hoop loading. (From Spencer & Hull 1978.)

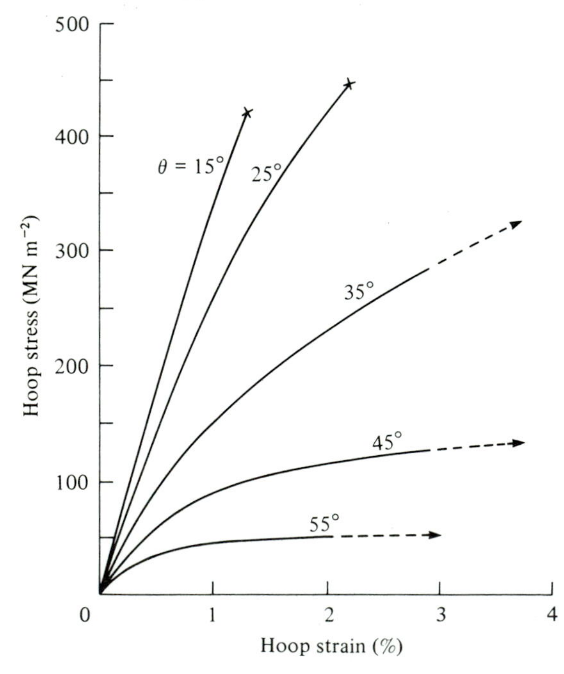

is due to viscoelastic flow of the resin matrix. At high stresses non-reversible deformation occurs and acoustic emission tests indicate the onset of cracking processes.

The prediction of the shape of the stress–strain curves in Fig. 8.12 requires a detailed knowledge of the viscoelastic properties of the resin and of the cracking processes in the laminae to estimate the degree of uncoupling and load transfer throughout the test. There is no adequate treatment at the present time. It might be argued that the laminate has failed at the onset of non-linearity and that the subsequent stress–strain response is not important. This is certainly not so in some applications.

The stresses in the individual laminae of an angle-ply tube tested in *biaxial tension* with $\sigma_H/\sigma_A = 2$, calculated from data in Fig. 6.17, are given in Fig. 8.13. Again, the exact positions of the curves depend on the properties of the laminae. The additional axial stress has a profound effect on the stresses in the laminae (cf. Fig. 8.11) and the stress–strain curves and the micromechanisms of laminate break-down are distinctly different. Transverse cracking dominates at all values of θ. The hoop stress–hoop strain curves of tubes wound at $\theta = 15°$, 35°, 45° and 55°

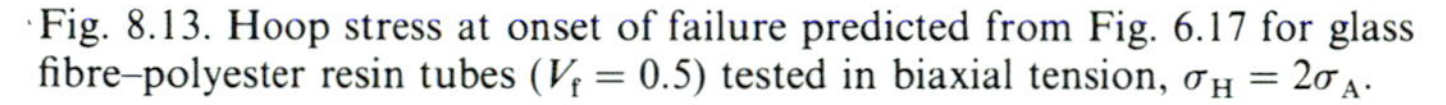

Fig. 8.13. Hoop stress at onset of failure predicted from Fig. 6.17 for glass fibre–polyester resin tubes ($V_f = 0.5$) tested in biaxial tension, $\sigma_H = 2\sigma_A$.

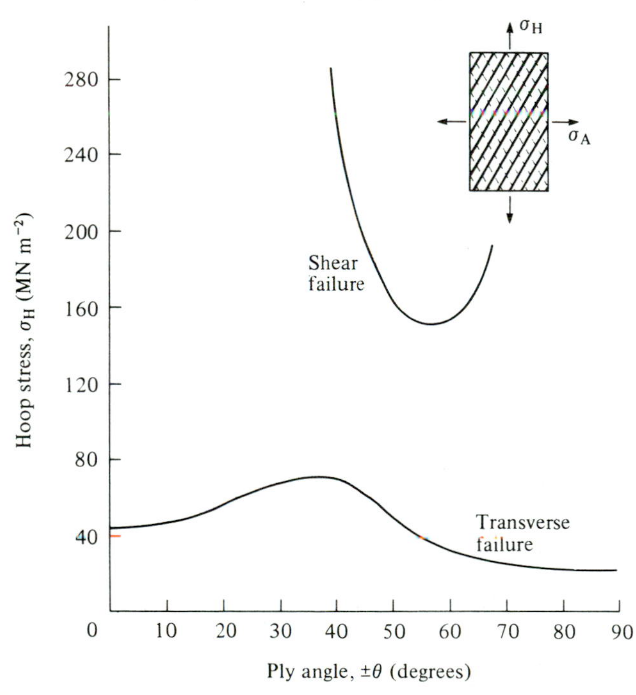

are shown in Fig. 8.14. In all these tests initial failure of the tubes is associated with transverse cracking as expected from the low transverse strength and the large values of $\sigma_\perp$.

The behaviour of the $\theta = 35°$ ($\phi = 55°$) tubes with $\sigma_H/\sigma_A = 2$ deserves special consideration because there are some important features which relate to the design of structures made from composite materials. One approach to the design of laminate structures is the use of 'netting analysis' in which it is assumed that the load bearing properties of the resin are negligible, or, alternatively, that the full load is taken by the fibres. This is, of course, a considerable over-simplification and laminate theory is to be preferred. By netting analysis $\sigma_\perp = \tau_\# = 0$. Introducing this into the transformation equations (6.31) for a single lamina gives

$$\left.\begin{aligned} \sigma_x &= \sigma_\parallel \cos^2 \phi \\ \sigma_y &= \sigma_\parallel \sin^2 \phi \\ \tau_{xy} &= \sigma_\parallel \sin \phi \cos \phi \end{aligned}\right\} \tag{8.10}$$

Fig. 8.14. Hoop stress–hoop strain curves of glass fibre–polyester resin angle-ply tubes ($V_f = 0.5$) tested in biaxial tension, $\sigma_H = 2\sigma_A$. (From Spencer & Hull 1978.)

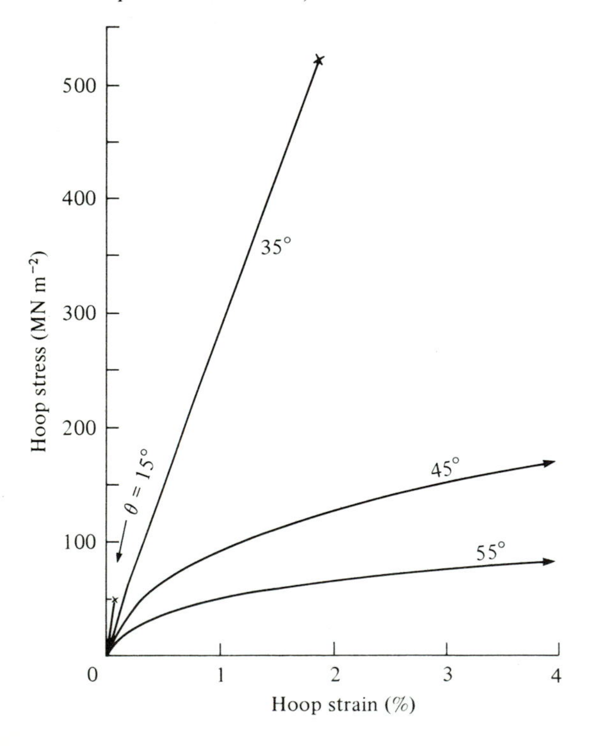

When $\sigma_y = \sigma_H$, $\sigma_x = \sigma_A$ and $\sigma_H/\sigma_A = 2$

$$\sigma_H/\sigma_A = \sigma_\parallel \sin^2\phi/\sigma_\parallel \cos^2\phi = 2 \tag{8.11}$$

or

$$\tan^2\phi = 2$$

Thus, there is only one angle, $\phi = 54.74°$, the so-called 'ideal' angle, which satisfies the requirements of netting analysis for the stress ratio $\sigma_H/\sigma_A = 2$. Netting analysis predicts that at all other angles for this stress ratio the load carrying capacity of the laminate is zero. When a load is applied the fibres will rotate towards the ideal angle.

The application of this approach to the results for tubes with $\phi = \pm 55°$, shown in Fig. 8.14, is illustrated in a more schematic form in Fig. 8.15. The predicted slope of the hoop stress–hoop strain and hoop stress–axial strain curves is obtained from netting analysis as follows. The stress parallel to the fibres is obtained from equation (8.10).

$$\sigma_{\parallel\,\mathrm{net}} = \sigma_H/\sin^2\phi = 1.5\,\sigma_H \tag{8.12}$$

Similarly, since the strain in the tube must be due entirely to extension of the fibres

$$\epsilon_{\parallel\,\mathrm{net}} = 1.5\,\sigma_H/V_f E_f \tag{8.13}$$

Fig. 8.15. Schematic representation of stress–strain response of $\phi = 55°$ ($\theta = 35°$) tube tested in biaxial tension, $\sigma_H = 2\sigma_A$.

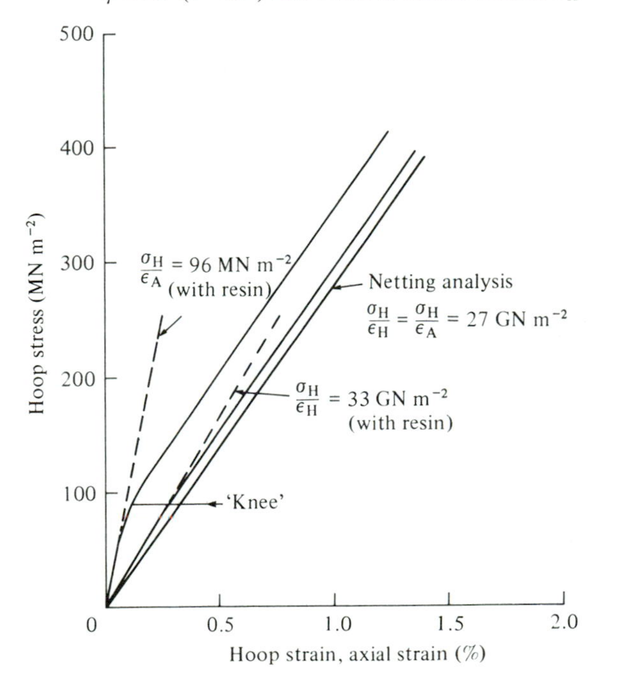

The axial and hoop strains will be the same as the strains in the fibres, thus

$$\epsilon_{\mathrm{H\,net}} = \epsilon_{\mathrm{A\,net}} = \epsilon_{\parallel\,\mathrm{net}} \tag{8.14}$$

and

$$\frac{\sigma_{\mathrm{H}}}{\epsilon_{\mathrm{H\,net}}} = \frac{\sigma_{\mathrm{H}}}{\epsilon_{\mathrm{A\,net}}} = \frac{V_{\mathrm{f}} E_{\mathrm{f}}}{1.5} \tag{8.15}$$

Taking $E_{\mathrm{f}} = 76$ GN m^{-2}, $V_{\mathrm{f}} = 0.53$ gives a slope $\sigma_{\mathrm{H}}/\epsilon_{\mathrm{H\,net}} =$ 26.9 GN m^{-2}.

The difference between the predictions of netting analysis and the experimental curves in Fig. 8.15 is due to the effect of the resin in binding the fibres together. Thus, the resin has a pronounced effect on the initial slope, particularly $\sigma_{\mathrm{H}}/\epsilon_{\mathrm{A}}$. During loading the applied stress results in resin shear and in fracture of the laminae. This leads to a progressive transfer of the load to the fibres and a change in slope towards that predicted by netting analysis. The experimentally measured slopes $\sigma_{\mathrm{H}}/\epsilon_{\mathrm{H}}$ and $\sigma_{\mathrm{H}}/\epsilon_{\mathrm{A}}$ are readily predicted by using laminate theory.

The explanation of the 'knee' in the curves in Fig. 8.15 is similar to that used for cross-ply laminates in Section 8.2. The first deviation from

Fig. 8.16. External appearance of a $\phi = 55°$ glass fibre–polyester resin tube tested in biaxial tension, $\sigma_{\mathrm{H}} = 2\sigma_{\mathrm{A}}$, to a hoop stress of 120 MN m^{-2}. White streaks are transverse cracks which form in all layers of the wall of the tube. Both long and short cracks are present and cracks intersect each other at interlaminar surfaces. (From Jones & Hull 1979.)

linearity is probably due to a small amount of resin creep but the major effect is associated with transverse cracking. Cracking can be detected by using acoustic emission and is clearly visible in the wall of the pipe (Fig. 8.16) and in polished sections. The cracks are regularly spaced and the average spacing decreases as the internal pressure increases in accordance with multiple cracking theory.

Final failure of the tube occurs when the internal pressure is sufficient to cause fibre fracture since eventually all the load is taken by the fibres. From equation (8.12) $\sigma_{\parallel} = 1.5\,\sigma_{H}$ so that a burst pressure of $\sigma_{H} = 500$ MN m^{-2} (see Fig. 8.14) corresponds to an effective fibre strength of 1.4 GN m^{-2} assuming $V_f = 0.53$. This is about 40% of the strength of freshly drawn glass fibres. The broken fibres at the site of burst are shown in Fig. 8.17. Experiments on tubes using internal pressure usually require the inside of the tube to be lined with a flexible rubber sleeve to prevent the pressurising fluid escaping through the transverse cracks created at lower pressures.

Fig. 8.17. Final failure region of a pipe tested in biaxial tension, $\sigma_H = 2\sigma_A$. A crack has formed parallel to one set of fibres and fibre fracture has occurred in other set. (From Hull, Legg & Spencer 1978.)

8.4 Edge effects in angle-ply laminates

As mentioned in the previous section, the presence of free edges can have a large effect on the failure processes of angle-ply laminates. This is important when analysing the results on finite width specimens and has practical applications in relation to design where free edges and holes are involved. The problem has been approached analytically by Pipes & Pagano (1970) who calculated the changes in stress distribution at laminate boundaries.

In the classical lamination theory introduced in Chapter 6 only the stresses in the plane of the laminate σ_x, σ_y and τ_{xy} are considered. No account is taken of interlaminar stresses σ_z, τ_{zx} and τ_{zy}; see

Fig. 8.18. Variation of interlaminar stresses across the width $2b$ of a four layer symmetric $\pm 45°$ laminate tested in uniaxial tension in x-direction. ($\theta = \pm 45°$). Calculations by Pipes & Pagano (1970) for a Type I carbon fibre–epoxy resin material using properties of individual laminae: $E_{\parallel} = 140$ GN m^{-2}, $E_{\gg} = 14.5$ GN m^{-2}, $G_{\parallel \gg} = 6.0$ GN m^{-2} and $\nu_{\parallel \gg} = 0.21$. (Width $2b$ is taken as 16 times thickness of individual laminae.)

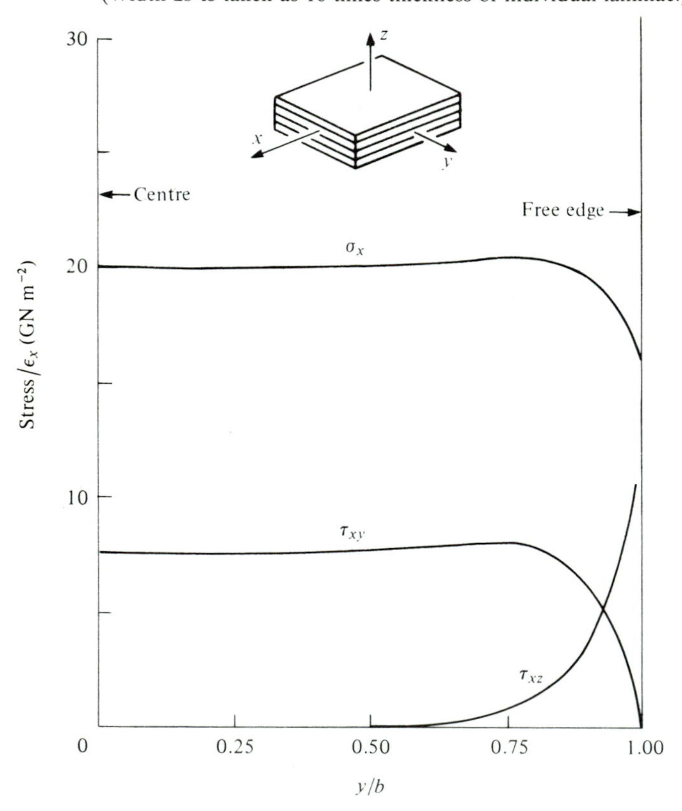

Fig. 8.19. Interlaminar shear stress as a function of ply angle for carbon fibre–epoxy resin laminate; see Fig. 8.18. (From Pipes & Pagano 1970.)

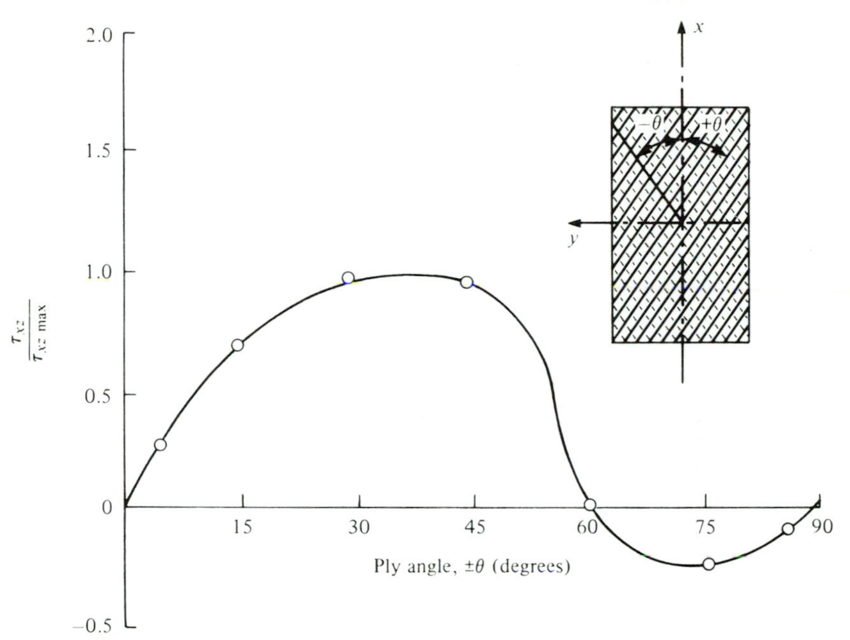

Fig. 8.20. Stress–strain curves of glass fibre–epoxy resin angle-ply laminates ($V_f = 0.60$). (After Rotem & Hashin 1975.)

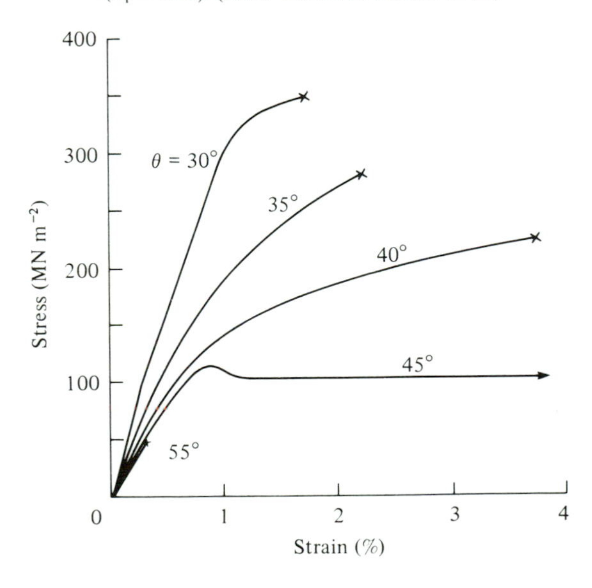

Fig. 8.18. In classical theory it is assumed that τ_{xy} is constant across the width whereas, in practice, τ_{xy} must reduce to zero at the free edges of the laminate. Pipes and Pagano have shown that additional out-of-plane stresses arise because of this requirement. The magnitude of these stresses depends on the elastic properties, orientation and stacking sequence of the laminae. Thus, for example, for an angle-ply laminate tested in tension in the x-direction the variation in σ_x, τ_{xy} and τ_{xz} across the width is shown in Fig. 8.18. τ_{xy} reduces to zero at the edge and σ_x decreases also whereas τ_{xz} increases. The edge effect extends into the laminate to a distance approximately equal to the thickness of the laminate. Although these results apply to a carbon fibre–epoxy resin laminate, similar effects are to be expected in other systems. Fig. 8.19 illustrates the influence of fibre orientation on the edge effect. The

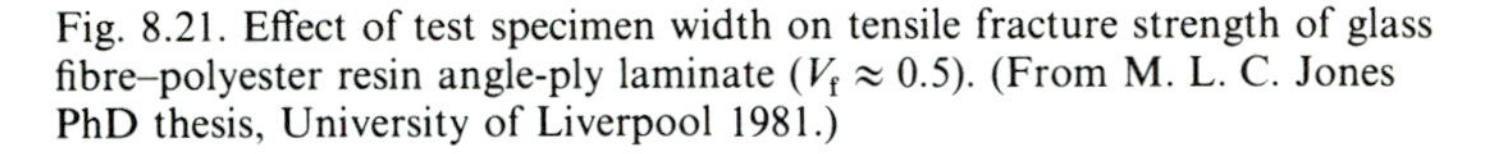

Fig. 8.21. Effect of test specimen width on tensile fracture strength of glass fibre–polyester resin angle-ply laminate ($V_f \approx 0.5$). (From M. L. C. Jones PhD thesis, University of Liverpool 1981.)

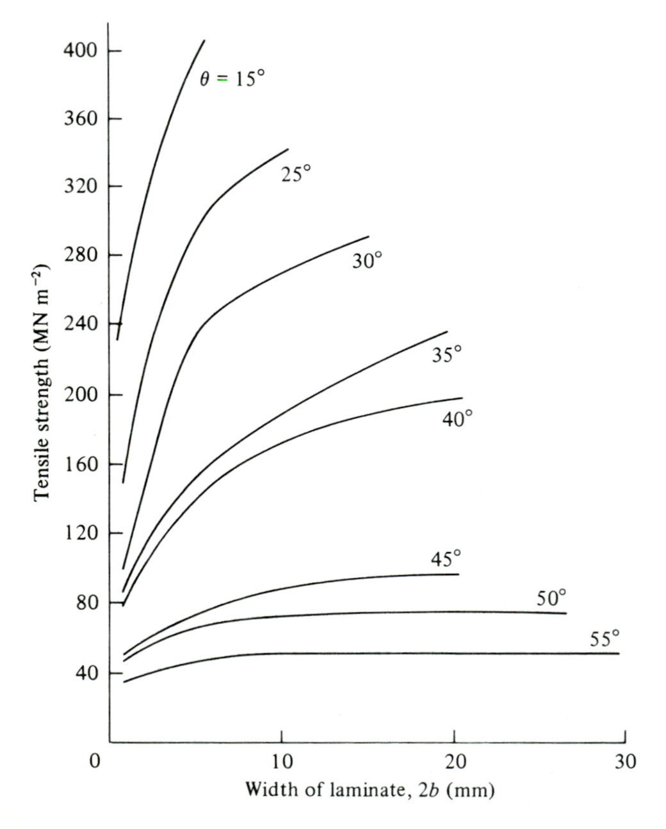

maximum interlaminar shear stresses occur at $\theta = 35°$ and the stresses are zero at 0°, 60° and 90°.

The edge effect can be demonstrated by using tensile tests on parallel sided angle-ply laminates. For wide specimens the stress–strain curves should be close to those observed in tube tests in pure hoop loading. An example is given in Fig. 8.20 from work on glass fibre–epoxy resin laminates by Rotem & Hashin (1975). They showed that the failure mechanisms are strongly dependent on θ and that the shape of the stress–strain curves can be predicted using a non-linear shear stress–shear strain response for the individual laminae. The very large displacement at $\theta = 45°$ is particularly interesting since it implies very large shear displacements. The curves in Fig. 8.20 are similar to those shown in Fig. 8.12 which were obtained from tube tests.

Fig. 8.22. Effect of test specimen width on tensile stress for transverse cracking and complete fracture in angle-ply laminates ($\theta = 50°$), for the same materials as in Fig. 8.21. (From M. L. C. Jones PhD thesis, University of Liverpool 1981.)

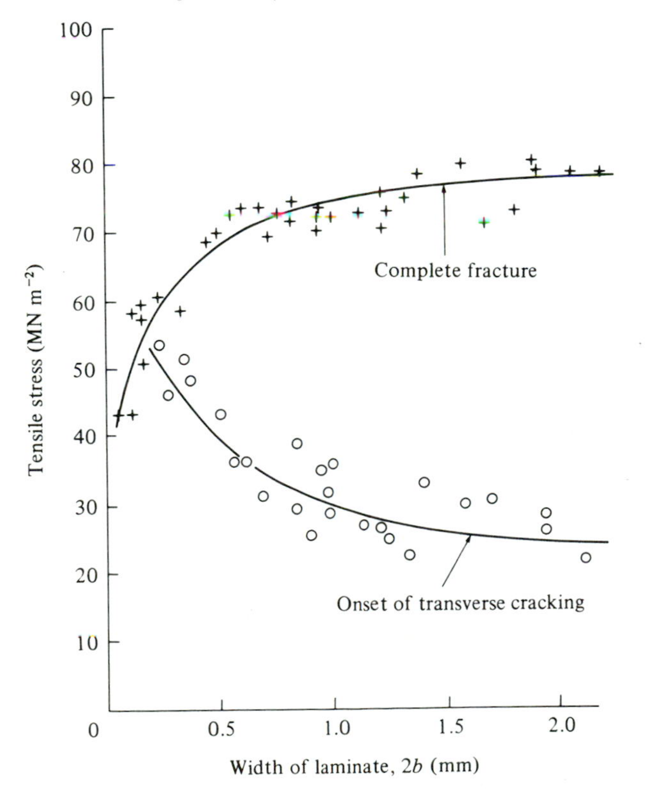

When tensile tests are carried out on specimens with different widths, i.e. $2b$ in Fig. 8.18, it is found that the final strength is strongly dependent on width as well as ply-angle as illustrated in Fig. 8.21. The width effect is more pronounced at low values of θ owing to the larger interlaminar shear stresses (Fig. 8.19). As the width increases, the tensile strength increases indicating that the effect of free edges decreases with increasing width. The edge effects are very pronounced in narrow specimens and there is a completely different stress distribution. Thus, for example, for $\theta = 50°$ laminate theory predicts that transverse cracking will dominate (see Fig. 8.11). Experimentally, it is found that as the width decreases the applied stress at the onset of transverse cracking increases and that below a critical width transverse cracking is suppressed (Fig. 8.22). Failure is then due entirely to interlaminar shear effects.

The high interlaminar shear stress τ_{xz} at the free edges can be understood in terms of the tendency of the individual laminae to rotate, as illustrated in Fig. 6.9, so that the fibres lie closer to the tensile axis.

Fig. 8.23. Schematic view of intralaminar and interlaminar shear associated with a tensile test on an angle-ply laminate.

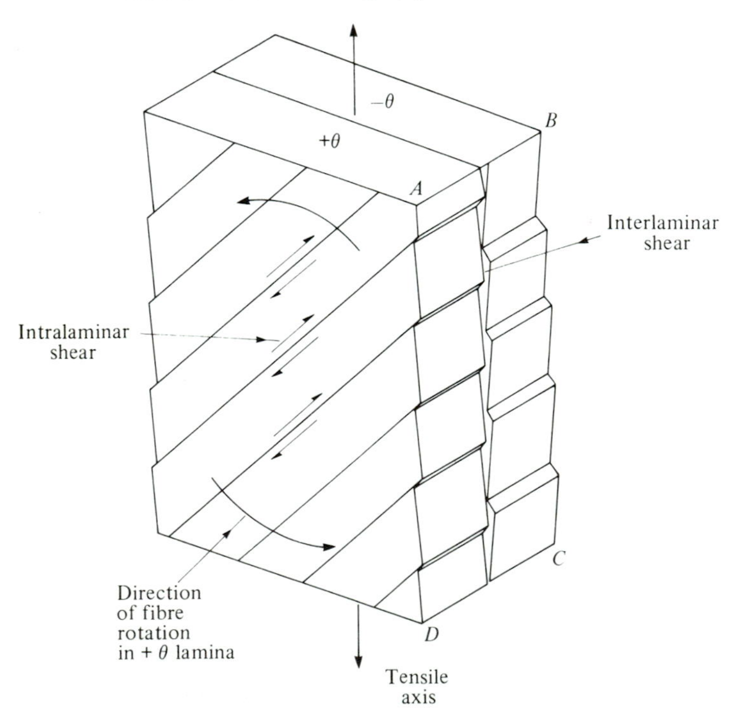

Fig. 8.24. Interlaminar crack at edge (*ABCD* in Fig. 8.23) of a glass fibre–polyester resin angle-ply laminate. (From M. L. C. Jones PhD thesis, University of Liverpool 1981.)

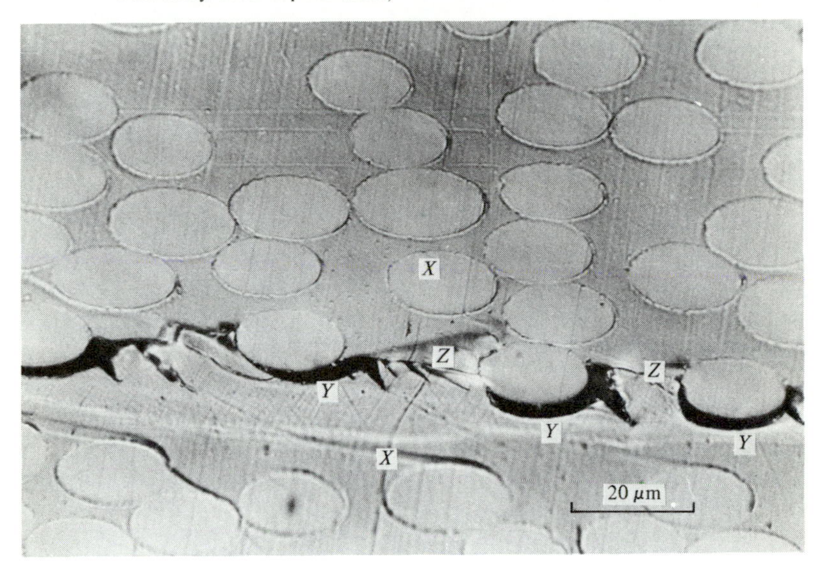

This means that adjacent laminae move in opposite directions and shear displacements or splitting must occur between the laminae (interlaminar). Similarly, shear must take place within each lamina so that rotation of the fibres can occur (intralaminar). These effects are represented schematically in Fig 8.23. The shear displacements and cracks can be studied by examining the free edges of the laminae, such as surface *ABCD* in Fig. 8.23. An example of an interlaminar crack is shown in Fig. 8.24. The large amount of resin shear is evident from the displacement of the fine polishing marks across *XX*. Debonding and resin shear fracture is evident at *Y* and *Z* respectively. The difference in focus indicates that out-of-plane displacements have occurred. An example of an intralaminar shear crack is shown in Fig. 7.36.

References and further reading

Amijuma, S. & Adachi, T. (1979) Non-linear stress–strain response of laminated composites. *J. Comp. Mater.* **13**, 206–18.

Aveston, J. & Kelly, A. (1973) Theory of multiple fracture of fibrous composites. *J. Mater. Sci.* **8**, 352–62.

Aveston, J. & Kelly, A. (1980) Tensile first cracking strain and strength of hybrid composites and laminates. *Phil. Trans. R. Soc. Lond.* A **294**, 519–34.

Ewins, P. D. & Potter, R. T. (1980) Some observations on the nature of fibre reinforced plastics and the implications for structural design. *Phil. Trans. R. Soc. Lond.* A **294**, 507–17.

Garrett, K. W. & Bailey, J. E. (1977*a*) Multiple transverse fracture in 90° cross-ply laminates of a glass fibre-reinforced polyester. *J. Mater. Sci.* **12**, 157–68.

Garrett, K. W. & Bailey, J. E. (1977*b*) The effect of resin failure strain on the tensile properties of glass fibre-reinforced polyester cross-ply laminates. *J. Mater. Sci.* **12**, 2189–94.

Greenwood, J. H. (1977) German work on GRP design *Composites* **8**, 175–84.

Hull, D., Legg, M. J. & Spencer, B. (1978) Failure of glass/polyester filament wound pipe. *Composites* **9**, 17–24.

Jones, M. L. C. & Hull, D. (1979) Microscopy of failure mechanisms in filament wound pipe. *J. Mater. Sci.* **14**, 165–74.

Ogorkiewicz, R. M. (1973) Orthotropic characteristics of glass fibre-epoxy laminates under plane stress. *J. Mech. Engng Sci.* **15**, 102–8.

Parvizi, A. & Bailey, J. E. (1978) On multiple transverse cracking in glass fibre epoxy cross-ply laminates *J. Mater. Sci.* **13**, 2131–6.

Pipes, R. B. & Pagano, N. J. (1970) Interlaminar stresses in composite laminates under axial extension. *J. Comp. Mater.* **4**, 538–48.

Puck, A. & Schneider, W. (1969) On failure mechanisms and failure criteria of filament-wound glass-fibre/resin composites. *Plast. Polym.* **37**, 33–43.

Rotem, A. & Hashin, S. (1975) Failure modes of angle-ply laminates. *J. Comp. Mater.* **9**, 191–206.

Rotem, A. (1977) The discrimination of micro-fracture modes of fibrous composite material by acoustic emission technique. *Fibre Sci. Technol.* **10**, 101–20.

Sims, G. D. (1976) Stress-wave emission from polymeric materials. *Plast. Rubber: Mater. Applics* **2**, 205–15.

Spencer, B. & Hull, D. (1978) Effect of winding angle on the failure of filament wound pipe. *Composites* **9**, 263–71.

9 Strength of short fibre composite materials

9.1 Introduction

The prediction and description of the strength properties has been restricted so far to continuous fibre composite materials with a thermosetting resin matrix. Even for the relatively simple fibre arrangements involved it is difficult to provide a complete account of the stress–strain response and the final fracture strength. The problems are more difficult and less amenable to rigorous analysis when the fibres are discontinuous. A full description of all the relevant strength properties is not yet possible and there is a dearth of information on these properties for well-characterised materials. Part of the difficulty is the large number of geometrical and material variables which are possible.

To obtain some insight into the problem three groups of short fibre materials are considered in this chapter. The ideas which have been developed for continuous aligned materials are applied to these materials with particular reference to tensile strength.

In the first group of materials the short fibres are aligned parallel to each other. This is by far the simplest configuration and it is possible to obtain composite materials in this form using wet lay-up techniques. Special methods are required to align the chopped fibres to produce a mat which is then infiltrated with resin. A high degree of alignment can also be obtained using injection moulding and extrusion processes by careful control of the flow geometry (see Section 4.9). The first part of this chapter deals with the properties of these rather ideal short fibre materials. The primary variables, in addition to matrix properties, are fibre length, fibre length distribution and interface strength.

The second group of materials contain in-plane randomly oriented long chopped fibres (chopped-strand mat, CSM). They are made by hand lay-up techniques with a thermosetting resin, usually a polyester. In normal practice the fibres are given a size treatment to hold them together in the strand so that the fibre bundles retain their integrity. Thus, the reinforcing 'unit' is the fibre bundle. A rather similar effect can be obtained in sheet moulding compound (SMC) and dough moulding compound (DMC) which are also based on thermosetting resins. The size on the fibre can be selected to obtain different amounts of fibre dispersion during moulding leading to different properties, particularly with respect to fracture toughness.

The fibre length in SMC and DMC is usually much smaller than in CSM. DMC differs from SMC in that it contains a large amount of inert filler, such as calcium carbonate, and the fibres have a three-dimensional random distribution before moulding. SMC has a two-dimensional random distribution of fibres before moulding. A special grade of SMC is obtained by aligning the fibres in the SMC 'pre-preg' to obtain strong in-plane directional properties. During moulding of SMC and DMC the flow processes lead to alignment of the fibres which varies from one part of the mould to another depending on the flow field.

The third group of materials to be considered are based on injection moulded products which have short fibres as the reinforcing element. The matrix is usually a thermoplastic although some thermosetting matrices are in use. The fibres may be completely random but usually show a degree of preferred orientation because of the flow fields. Large differences in orientation may occur from one part of a moulding to another.

A direct consequence of the more random distribution of fibres is the lower values of V_f in these materials and large resin rich areas may occur. The lower V_f is due to less efficient packing of the fibres and limitations in the moulding process. Thus, for example, the viscosity of a thermoplastic melt increases when fibres are added so that high injection pressures and temperatures have to be used which may be greater than the practical limits of the equipment available.

9.2 Short aligned fibres

In Section 7.2 it was shown that the strength of continuous aligned fibre composite materials parallel to the fibres, $\sigma_{\parallel}^*$, can be described in terms of the fibre and matrix strengths and that the dependence of strength on V_f is related to the relative values of ϵ_f^* and ϵ_m^*. A similar pattern of behaviour is expected in short aligned fibre composite materials but the efficiency of strengthening by the fibres will depend on fibre length. Consider, for example, a high V_f material in which $\epsilon_m^* > \epsilon_f^*$. For continuous fibres equation (7.7) holds, i.e.

$$\sigma_{\parallel}^* = \sigma_f^* V_f + \sigma_m' (1 - V_f) \tag{9.1}$$

It is assumed that the strength is limited by the fracture strength of the fibres and that once σ_f^* is reached complete fracture occurs.

To apply this approach to material containing short fibres it is necessary to relate the stress in the fibres to the applied stress by means of the analysis discussed in Sections 5.4 and 7.3. Assuming a simple relation which gives a linear build-up of stress from the fibre ends (see Fig. 7.17) the critical length of fibre, defined as the minimum length of fibre required for the stress to build up to the fracture strength of the fibre, σ_f^*, is given by

$$l_c = r\sigma_f^*/\tau \tag{9.2}$$

where $2r$ is the diameter of the fibre and τ is the shear stress parallel to the fibre resisting pull-out. This is related either to the shear strength of the matrix or the strength of the fibre–matrix interface. Alternatively, it may be assumed that the fibre is elastically bonded to the matrix.

Consider an aligned short fibre composite material as illustrated in Fig. 9.1 with fibre length l. Two different failure conditions are possible depending on the relative values of τ, σ_f^* and σ_m^*. Suppose, firstly, that fibre fracture occurs before matrix fracture. The build-up in stress in the fibres, as the load on the composite is increased, can be represented by Fig. 9.2*a*. When the stress for fibre fracture is reached the average stress in the fibre is

$$\bar{\sigma}^* = \left[\frac{(l-l_c)+\frac{1}{2}l_c}{l}\right]\sigma_f^* \tag{9.3}$$

$$\bar{\sigma}^* = (1-l_c/2l)\,\sigma_f^*$$

Substituting equation (9.2) gives

$$\bar{\sigma}^* = (1-\sigma_f^*\,r/2l\tau)\,\sigma_f^* \tag{9.4}$$

and combining these equations with equation (9.1) by substituting $\bar{\sigma}^*$ for σ_f^* gives

$$\sigma_\parallel^* = (1-l_c/2l)\,\sigma_f^*\;V_f+\sigma_m'\;(1-V_f)$$

or

$$\sigma_\parallel^* = (1-\sigma_f^*\,r/2\tau l)\,\sigma_f^*\;V_f+\sigma_m'\,(1-V_f) \tag{9.5}$$

The second failure condition, illustrated in Fig. 9.2*b*, occurs when the build-up of stress in the fibre is insufficient to cause fibre fracture.

Fig. 9.1. Schematic representation of a section through an aligned short fibre composite material.

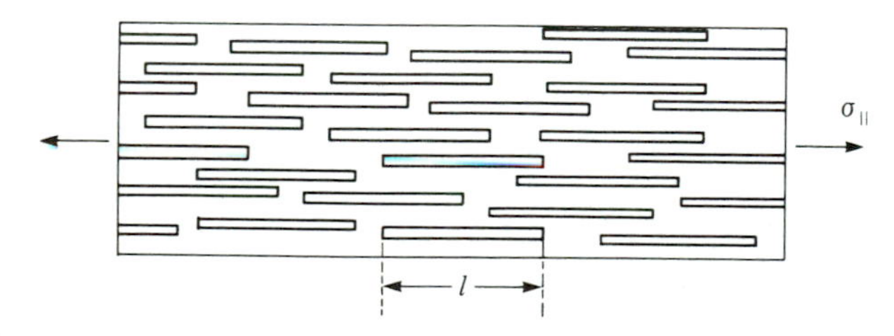

Fracture occurs when the matrix stress is reached and the average stress in the fibres is

$$\bar{\sigma}_f = l\tau/2r \tag{9.6}$$

assuming $l < l_c$. The strength of the composite material is given by

$$\sigma_{\parallel}^* = (l\tau/2r)\,V_f + \sigma_m^*(1-V_f) \tag{9.7}$$

Equations (9.5) and (9.7) can be used to predict the variation in strength with fibre length if σ_f^*, σ_m^* and l_c are known. Some results for a glass fibre–epoxy resin system are given in Fig. 9.3. For this system $\sigma_f^* = 1.8$ GN m^{-2}, $\sigma_m^* = 91.5$ MN m^{-2}, $d = 2r = 12.7$ µm and $V_f = 0.26$. The experimental points and the line predicted by using the equations are both shown. Since the value of l_c is not known the theoretical curve is fitted to the experimental results using different values of l_c. It is found that the best fit is obtained with $l_c = 12.7$ mm.

It follows from equation (9.2) that l_c is dependent on τ and hence the

Fig. 9.2. Change in stress distribution along fibre as load on composite material is increased (i) → (ii) → (iii). (*a*) Failure due to fibre fracture. (*b*) Failure due to matrix fracture.

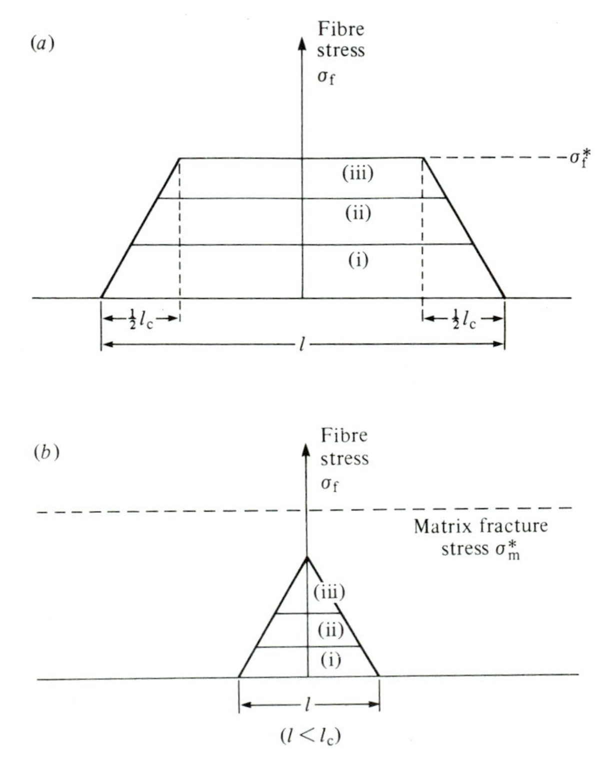

strength of the composite material will depend, as expected, on the strength of the interface or the shear strength of the matrix. For the system shown in Fig. 9.3 the strength of the discontinuous fibre material reaches the strength of the continuous fibre material when the fibre length is approximately 10 l_c and is 90% of the strength of 4 l_c. Similar experiments have been reported on carbon fibre–epoxy resin composite materials (see Dingle 1974).

When fibre alignment is produced by resin flow as in the injection moulding of fibre-filled thermoplastics there is considerable fibre breakdown during the processing operations. The fibre length is not constant and may show very wide variations. If all the fibres are shorter than l_c, and fracture occurs when the matrix fractures, the fracture strength is obtained from equation (9.7) and is given by

$$\sigma_{\|}^* = \sum_{l_i < l_c} \frac{l_i \tau}{2r} V_f + \sigma_m^*(1 - V_f) \tag{9.8}$$

If some of the fibres are longer than l_c and some shorter, combining equations (9.5) and (9.7) gives

$$\sigma_{\|}^* = \sum_{l_i < l_c} \frac{l_i \tau}{2r} V_i + \sum_{l_j > l_c} \left(1 - \frac{l_c}{2l_j}\right) \sigma_f^* V_j + \sigma_m'(1 - V_f) \tag{9.9}$$

where

$$\Sigma_i V_i + \Sigma_j V_j = V_f$$

provided fibre fracture occurs before matrix fracture. It is assumed in

Fig. 9.3. Effect of fibre length on strength of aligned glass fibre–epoxy resin composite material (From Hancock & Cuthbertson 1970.)

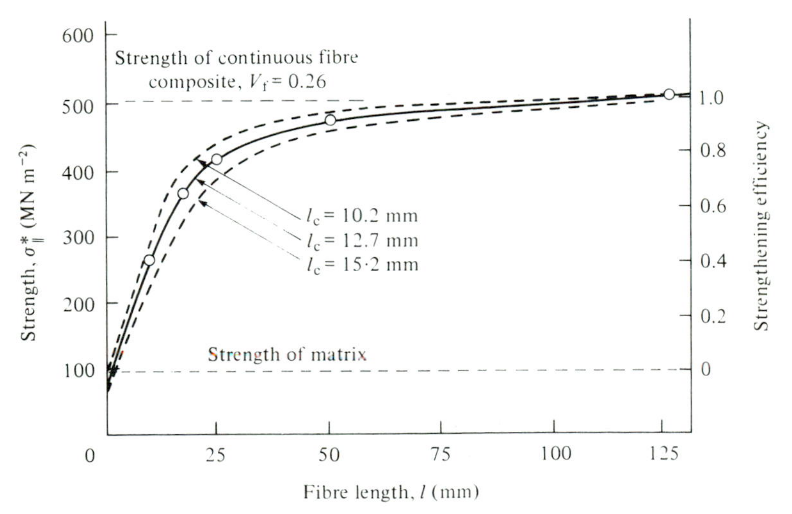

this simple approach that shrinkage stresses and differential contraction stresses play no part in determining the shear strength of the interface or the tensile strength of the matrix. These effects are accounted for in more detailed calculations.

Equation (9.9) indicates that, provided the fibre length distribution is constant, the longitudinal tensile strength will increase linearly with V_f. Some experimental results in support of this conclusion are shown in Fig. 9.4 for glass fibres in a nylon 6 matrix tested over a range of temperatures. At temperatures above 0 °C the matrix ($V_f = 0$) yields before fracture and at temperatures above 70 °C yielding occurs for all values of V_f used. The yield strength increases linearly with V_f. At 23 °C the addition of 5% fibres results in a change to brittle behaviour. The increase in strength with decreasing temperature is due to the increase in the strength of the matrix. The change in slope is indicative of an increase in the value τ, which is also related to the strength of the matrix. The reduction in slope at higher values of V_f is attributed to a reduction

Fig. 9.4. Effect of V_f and temperature on the yield (open symbols) and fracture (closed symbols) strengths of glass-filled nylon 6. (From Ramsteiner & Theysohn 1979.)

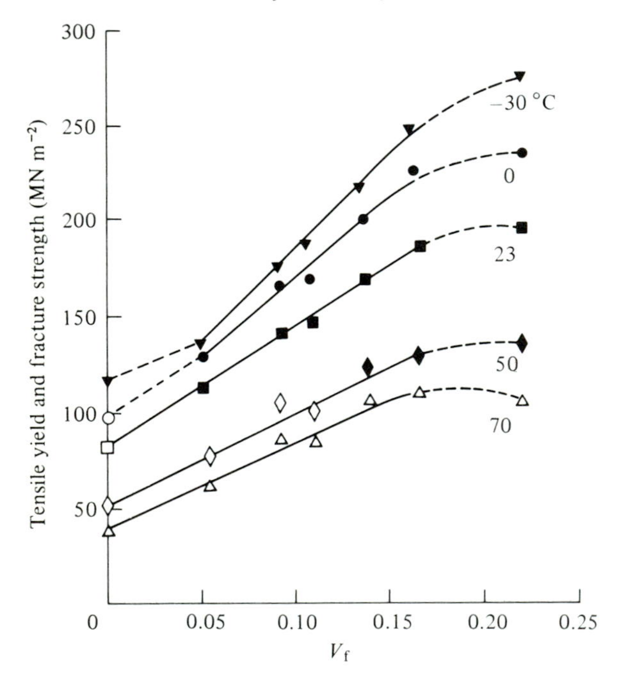

in fibre length due to fibre fracture following fibre contact during injection moulding. The reinforcing efficiency, defined by

$$k_1 = \frac{\sigma_{\parallel}^*(\text{short}) - \sigma_m'(1 - V_f)}{\sigma_{\parallel}^*(\text{continuous}) - \sigma_m'(1 - V_f)} \tag{9.10}$$

increases with decreasing temperature.

One of the consequences of a distribution of fibre lengths is that the slope of the stress–strain curve decreases with increasing strain even for material with elastically bonded fibres in a non-creeping matrix. This arises because the load carrying capacity and the efficiency of the fibres decreases as the strain increases (see Fig. 9.2). At low stresses and strains all the fibres make a contribution according to equation (9.3). As the strain increases a progressively smaller proportion of the fibres will reinforce according to equation (9.3) and an increasing proportion will follow equation (9.6). By expressing these equations in terms of the overall strain, it can be shown that the 'elastic' slope decreases as the strain increases. A typical curve is illustrated in Fig. 9.5 for a nylon 6.6–glass fibre composite material. Not all the non-linear behaviour

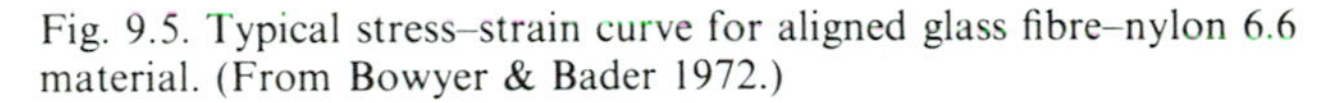

Fig. 9.5. Typical stress–strain curve for aligned glass fibre–nylon 6.6 material. (From Bowyer & Bader 1972.)

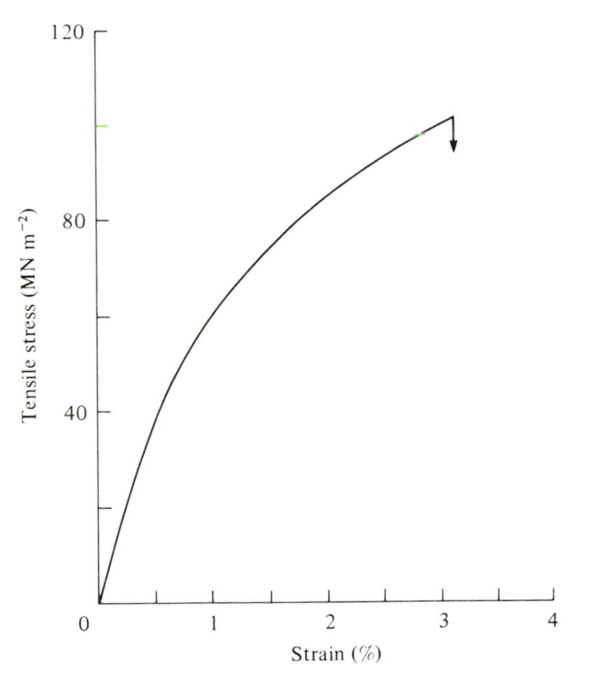

is due to elastic unloading of the fibres since resin creep occurs also.

The fracture surface has a characteristic appearance as shown in Fig. 9.6. For materials which fracture according to the conditions defined for equation (9.8), i.e. all the fibres are shorter than l_c at the fracture stress, the maximum length of fibre protruding from the fractured matrix is half the maximum length of the fibres in the unfractured material. In materials with $l > l_c$ some fibre break-down occurs before complete fracture. Thus, during crack propagation some fibres fracture in the plane of the crack and others fracture away from the crack plane, and final separation occurs by fibre pull-out.

The work of fracture depends on the amount of pull-out and on V_f.

Fig. 9.6. Scanning electron micrograph of the fracture surface of an aligned short fibre material, glass fibre–nylon ($V_f \approx 0.09$). (From Ramsteiner & Theysohn 1979.)

According to Kelly and co-workers (see Kelly 1970), the average work done per fibre is

$$W_{\mathrm{p}} = \tfrac{1}{12}\pi r \tau l^2 \quad (l \leqslant l_{\mathrm{c}}) \tag{9.11}$$

assuming that all the fibres pull-out so that the pull-out length various between 0 and $\frac{1}{2}l$. If only a fraction (l_{c}/l) of the fibres pull-out the average work done per fibre is

$$W_{\mathrm{p}} = (l_{\mathrm{c}}/l)\tfrac{1}{12}\pi r \tau l_{\mathrm{c}}^2 \quad (l \geqslant l_{\mathrm{c}}) \tag{9.12}$$

A schematic representation of the variation of the work of fracture due to pull-out with the fibre length l is shown in Fig. 9.7. For small l the work of fracture increases with fibre length because the pull-out length is increasing. It reaches a maximum at l_{c} and then drops rapidly because fibre fracture occurs for $l > l_{\mathrm{c}}$ and the amount of pull-out decreases.

The tensile strength of aligned discontinuous fibre materials varies with fibre angle in a similar way to that described in Section 7.8 for continuous fibres and the same arguments about failure criteria apply. Fracture can be interpreted in terms of longitudinal, transverse and shear type failures although there are few experimental data available. An example of the orientation dependence of the fracture strength of a glass fibre–polymethylmethacrylate material, $V_{\mathrm{f}} = 0.15$, is shown in Fig. 9.8. There is good agreement with the Tsai–Hill criterion (equation (7.34)). The values of $\sigma_{\perp}^*$ and $\tau_{\#}^*$ for discontinuous fibre material will be much less affected by fibre length than the corresponding values of $\sigma_{\parallel}^*$.

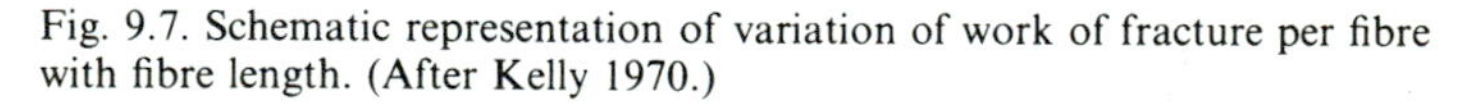

Fig. 9.7. Schematic representation of variation of work of fracture per fibre with fibre length. (After Kelly 1970.)

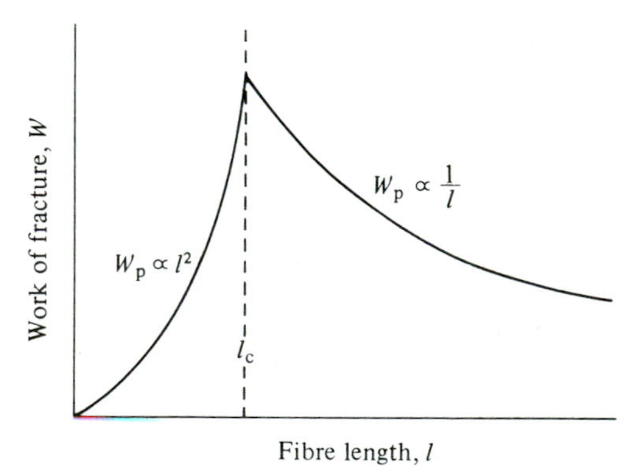

9.3 In-plane random fibres

As mentioned in the introduction (Section 9.1) the fibres may be held together in bundles or dispersed as individual fibres. In CSM processing the bundle integrity is maintained and little fibre fracture occurs after the fibres have been chopped to the pre-determined length. In SMC processing some fibre degradation and dispersion occurs depending on the size treatment given to the fibres. When the bundles are strongly bound the bundles act as the reinforcing unit. Thus, the reinforcing efficiency, fracture characteristics and work of fracture will depend on the degree of dispersion. This section is concerned primarily

Fig. 9.8. Tensile strength of aligned short fibre material as a function of angle. Glass fibre–PMMA ($V_f = 0.15$). Theoretical lines calculated using $\sigma^*_\parallel = 121$ MN m^{-2}, $\sigma^*_\perp = 60$ MN m^{-2} and $\tau^*_\# = 45$ MN m^{-2}. (From Ramsteiner & Theysohn 1979.)

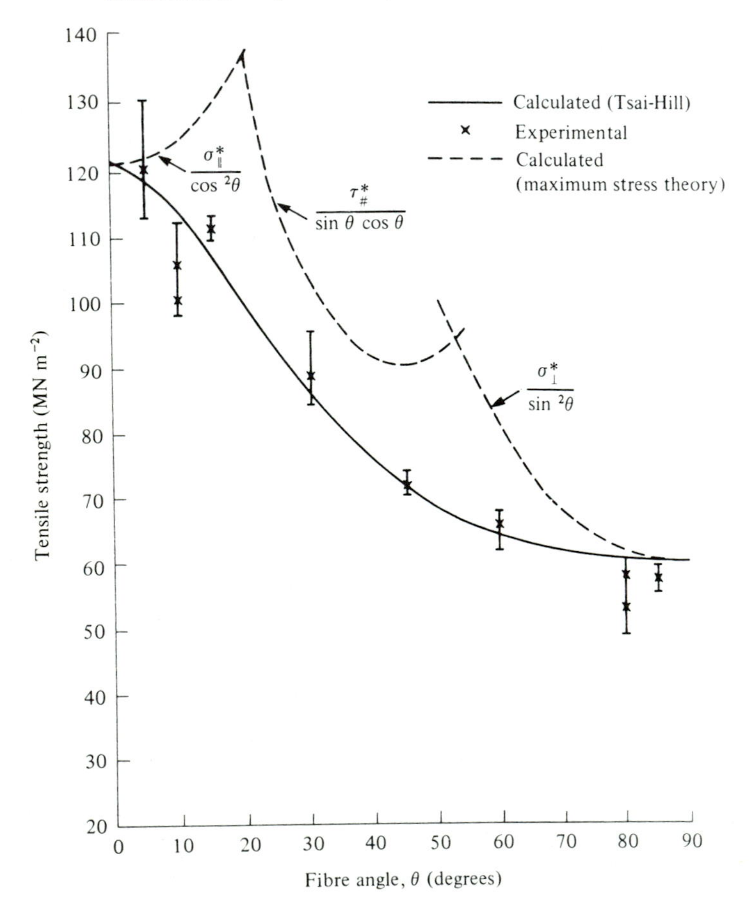

with materials in which the fibres are held in tightly packed bundles. An example of the microstructure is shown in Fig. 9.9. Each bundle has an approximately elliptical cross-section and contains typically about 400 fibres.

Since the reinforcing units are distributed randomly in the plane of sheet material it follows that when a test sample is tested in uniaxial tension some of the units will be parallel to the stress direction and others perpendicular to it. It is reasonable to assume that the strain is uniform across the width of the sheet so that the load bearing capacity and local failure mechanisms will depend on the orientation of each reinforcing unit. Failure of bundles oriented normal and parallel to the applied stress can be considered in terms of the behaviour of unidirectional laminae described in Chapter 7. Since $\epsilon_{\parallel}^{*} > \epsilon_{\perp}^{*}$ failure of the transversely oriented bundles will occur first. Fibre fracture in the longitudinally oriented bundles is unlikely since pull-out will occur before the critical stress is reached. Each bundle of 400 fibres has an effective aspect ratio of approximately $40r/L$ where L is the length of the bundle. This compares with an aspect ratio of $2r/L$ for single fibres of the same length. Thus, assuming the interface properties of the

Fig. 9.9. Polished section through a glass fibre–polyester resin chopped-strand mat laminate showing fibre bundles and resin rich regions.

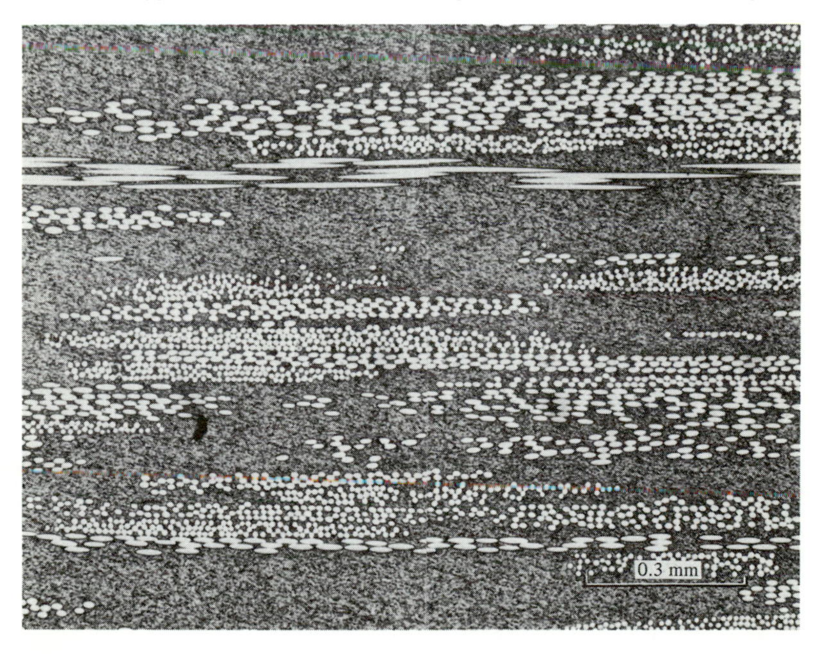

surface of the bundle are the same as the surface of the individual fibres, pull-out of bundles occurs at much lower stresses (cf. equation (9.2)).

A typical load–extension curve of a CSM polyester–glass laminate is shown in Fig. 9.10. The polyester resin used had a strain to failure of about 2% and the volume fraction of fibres was 0.17. Failure is progressive but three stages can be identified. The material behaves in a linear elastic manner to a strain of 0.3%. Deviation of the curve is due to the onset of cracking in the transversely oriented fibre bundles as illustrated in Fig. 9.11*a*. Cracking occurs preferentially within the bundle because the fibres are closely spaced giving high strain magnifications between the fibres, as described in Section 5.2. The cracking is readily detected by acoustic emission techniques.

As the strain increases, cracking occurs in bundles oriented at other angles to the tensile axis as predicted from the orientation dependence of the strength of unidirectional laminae (Section 7.8). The more pronounced change in slope, the 'knee', at a strain of 0.7% is due to the onset of resin cracking between the fibre bundles as shown in Fig. 9.11*b*. Further debonding, shear cracking and resin fracture occurs as the load increases until ultimate failure by complete separation at a strain of 1.5–2.0%.

There is no entirely satisfactory theory to account for the strength of in-plane random fibre materials. One approach is to use an empirical method to predict strength similar to that used to represent the V_f

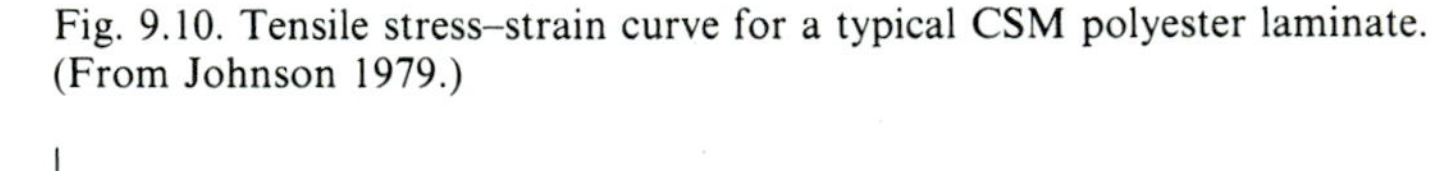

Fig. 9.10. Tensile stress–strain curve for a typical CSM polyester laminate. (From Johnson 1979.)

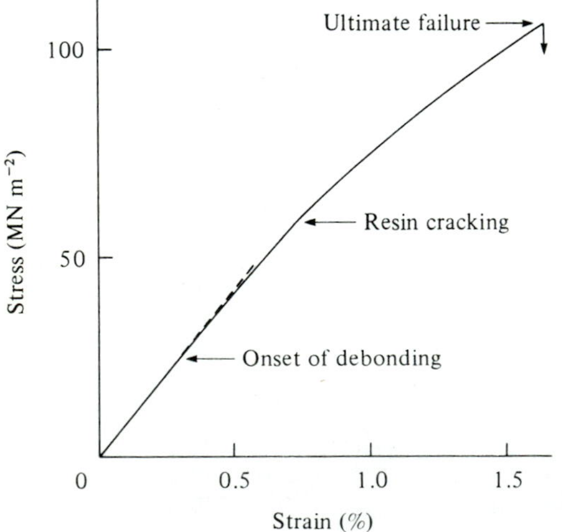

Fig. 9.11. Microcracking during tensile loading of a CSM glass fibre–polyester laminate. (*a*) Initial debonding in fibre bundles. (*b*) Resin cracking between fibre bundles. (From Owen 1970.)

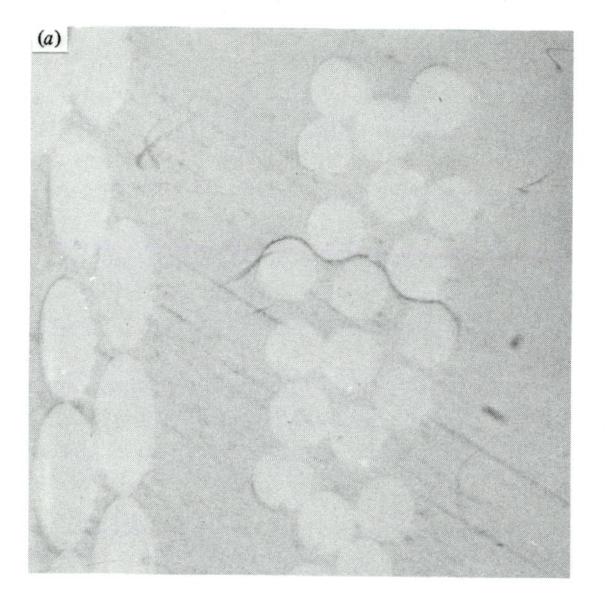

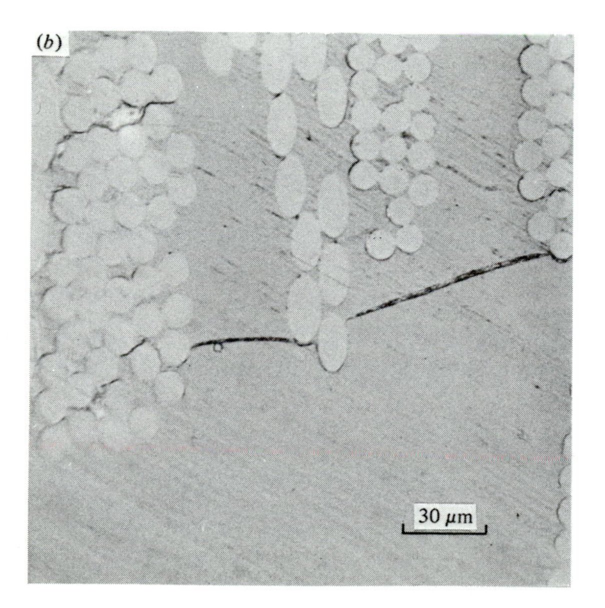

dependence of modulus in equation (5.40). Thus, the tensile strength of a material for which $\epsilon_f^* > \epsilon_m^*$ can be represented by a modified 'rule of mixtures' equation

$$\sigma_r^* = K_1 K_2 \sigma_f^* V_f \tag{9.13}$$

where K_1 is the strengthening efficiency of short fibres (cf. equation (9.10)) and K_2 is the orientation efficiency (subscript r stands for 'random'). There are very few data available to evaluate this relation. One example is given in Fig. 9.12 for an SMC material over a fairly narrow range of volume fractions. With a value of $\sigma_f^* = 1.8$ GN m^{-2} the data can be fitted to equation (9.13) using $K_1 = 0.8$ and $K_2 = 0.375$. The difficulty of establishing the values of these parameters in practice highlights the limitations of using equations of this form for the prediction of strength in unknown systems.

Two other methods of predicting the strength are worth mentioning since they involve some physical model for the composite material. In the first method the in-plane random material is treated as a stack of infinitesimally thin unidirectional laminae bonded together at different

Fig. 9.12. Variation of tensile strength with V_f for a glass fibre–polyester sheet moulding compound. $K_1 = 0.8$, $K_2 = 0.375$, $\sigma_f^* = 1.8$ GN m^{-2}. (Data from D. Pennington PhD thesis, University of Liverpool 1979.)

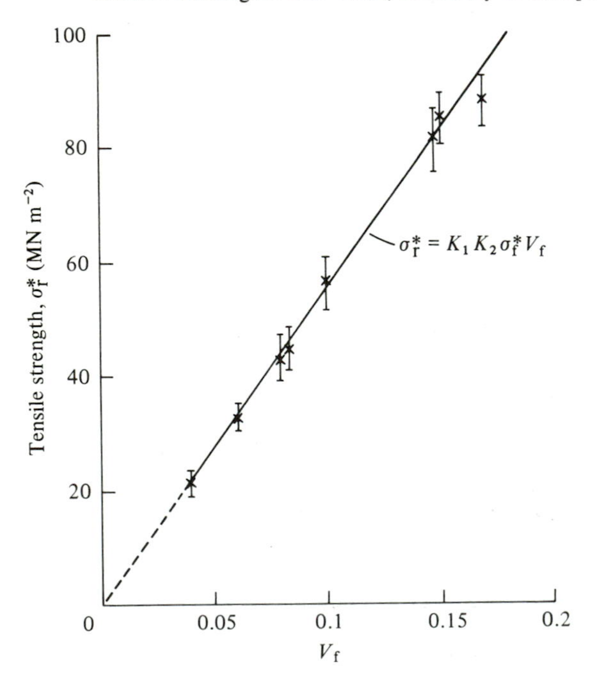

angles with a random angular distribution. The strength of each lamina is assumed to be given by the maximum stress criterion so that

$$\left.\begin{aligned} \sigma_\theta(\text{fibre failure}) &= K_1\sigma_\parallel^*/\cos^2\theta \\ \sigma_\theta(\text{shear failure}) &= \tau_\#^*/\sin\theta\cos\theta \\ \sigma_\theta(\text{transverse failure}) &= \sigma_\perp^*/\sin^2\theta \end{aligned}\right\} \tag{9.14}$$

where K_1 is the strength efficiency factor for the fibre length used. It is assumed that $\tau_\#^*$ and $\sigma_\perp^*$ are unaffected by fibre length. The strength of randomly oriented material is obtained by averaging the off-axis strength σ_θ over the range 0°–90° assuming that the ultimate failure strain is the same throughout the material. The resulting equation obtained by Chen (1971) is

$$\sigma_r^* = \frac{2\tau_\#^*}{\pi}\left[2 + \ln\left(\frac{K_1\sigma_\parallel^*\sigma_\perp^*}{\tau_\#^{*2}}\right)\right] \tag{9.15}$$

The dependence of strength on V_f is introduced through the effect of V_f on the properties of the individual laminae.

In the second method due to Halpin & Kardos (1978) the random in-plane material is modelled by a quasi-isotropic laminate consisting of a stack of unidirectional laminae oriented at 0°, 90° and ±45° as

Fig. 9.13. Stress–strain curve of a quasi-isotropic laminate (0°, 90°, ±45°). Glass fibre–epoxy resin, $V_f \approx 0.60$. (From Halpin & Kardos 1978.)

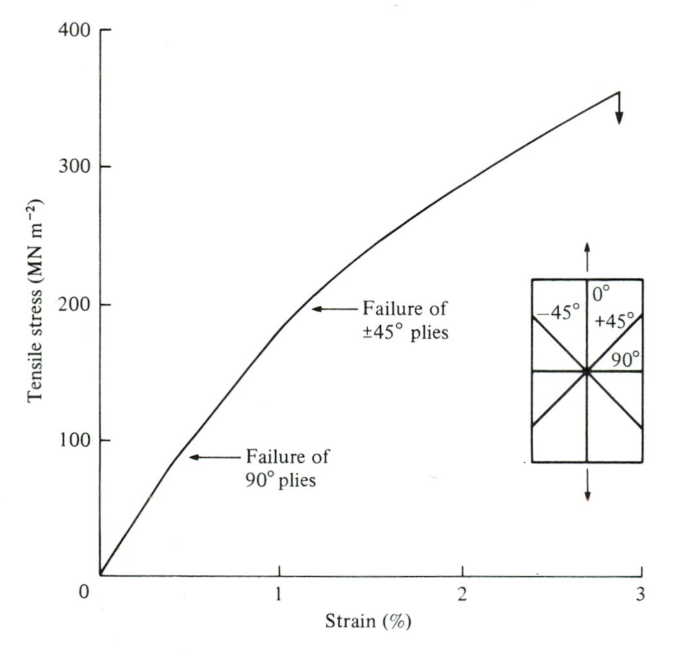

illustrated in Fig. 9.13. The failure of the laminate occurs by sequential failure of the laminae in a similar way to cross-ply laminates (see Section 8.2) and a typical example is shown in Fig. 9.13. The laminate strength is obtained by the iterative process of recalculating the stress to failure of residual laminae as each lamina fails. Using a maximum strain criterion for fracture Halpin and Kardos demonstrated good agreement between this model and some experimental results from an in-plane random glass fibre–epoxy resin laminate.

Both these methods are difficult to apply because it is necessary to have data on the failure strengths and strains of unidirectional laminae at a range of off-axis angles as a function of V_f and fibre length. Thus, effective prediction of in-plane random properties requires a large amount of preliminary testing of well-characterised materials.

Fig. 9.14. Static test results for chopped-strand mat–polyester resin tubes ($V_f \approx 0.20$). R = axial stress/hoop stress. (From Owen & Found 1972.)

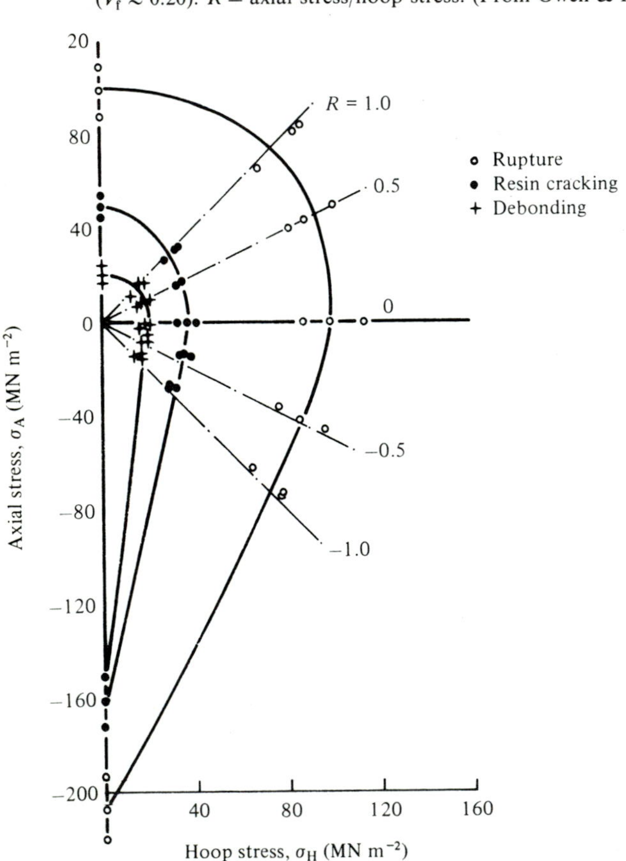

The failure of in-plane random material under more complex loading conditions can be understood by using the ideas and models already described although it must be emphasised that little systematic work has been done. One example will suffice from tests on chopped-strand mat laminates by Owen & Found (1972). The biaxial test results in Fig. 9.14 were obtained by using silicone rubber-lined tubes which were tested in combined internal pressure and axial load conditions. The stresses at the onset of debonding, resin cracking and final rupture (cf. Fig. 9.10) were measured at a number of principal stress ratios ($\sigma_A/\sigma_H = R$). The pure hoop loading ($R = 0$) and pure axial loading ($R = \infty$) correspond to the uniaxial test data described earlier. The failure stresses in compression loading are much greater than in tension. Pure shear loading occurs at $R = -1.0$.

Finally, it should be noted that in-plane random materials are isotropic in the plane of the fibres. By controlling the fibre orientation, specific strength properties can be designed into the material.

9.4 Injection moulded short fibre materials

These materials are usually made with thermoplastic polymer matrices which may be amorphous or semicrystalline (see Section 2.6). In the injection moulding process the feed-stock of intimately mixed fibres and polymer is heated in the feed-barrel to melt the polymer and is then injected into a mould. The molten polymer undergoes large shear and extensional flow which may lead to molecular orientation in the final product if it is 'frozen in'. The crystallisation of the polymer may also be affected by the melt flow and by the presence of the fibres. For example, epitaxial crystallisation of polymer molecules has been observed around carbon fibres. Thus, it is not always correct to assume that the matrix is isotropic and homogeneous.

The matrix effects are usually insignificant compared with the effects of fibre orientation. Melt flow during mould filling results in orientation of the fibres (see Section 4.9) which varies from one part of the material to another depending on the way in which mould filling occurs. Thus, the orientation distribution depends on the position of the gate or injection point in the mould and on the process conditions since these affect the flow properties of the fibre-filled melt. Fully three-dimensional random distributions of fibres are not expected and because of the orientation distribution variation within the material it is impossible to obtain experimental data which are entirely characteristic of random fibre material. The problem is illustrated in Fig. 9.15 for a very simple mould geometry, a flat rectangular plaque, with the gate in the centre of one of the edges. Provided the mould filling is well controlled, the

Fig. 9.15. Schematic representation of successive positions of melt front during injection moulding of a square plaque from a gate in the centre of one edge.

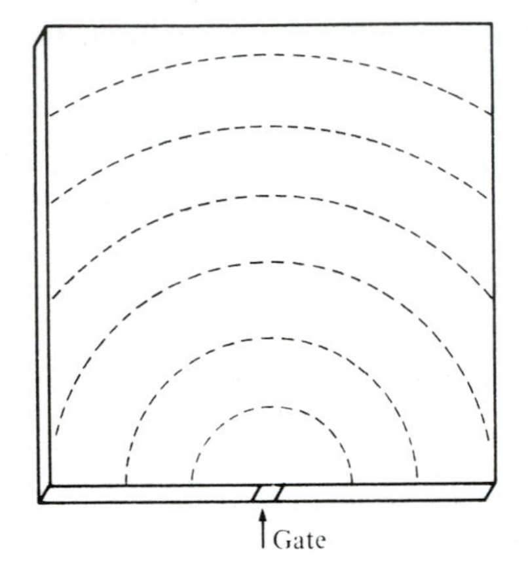

Fig. 9.16. Tensile stress–strain curves for injection moulded glass-filled nylon 6.6 ($V_f = 0.17$) with and without coupling agents applied to the glass fibre. Dry samples vacuum dried at 100 °C to constant weight. Wet samples immersed in boiling water to constant moisture uptake. (From Berry & Stanford 1977.)

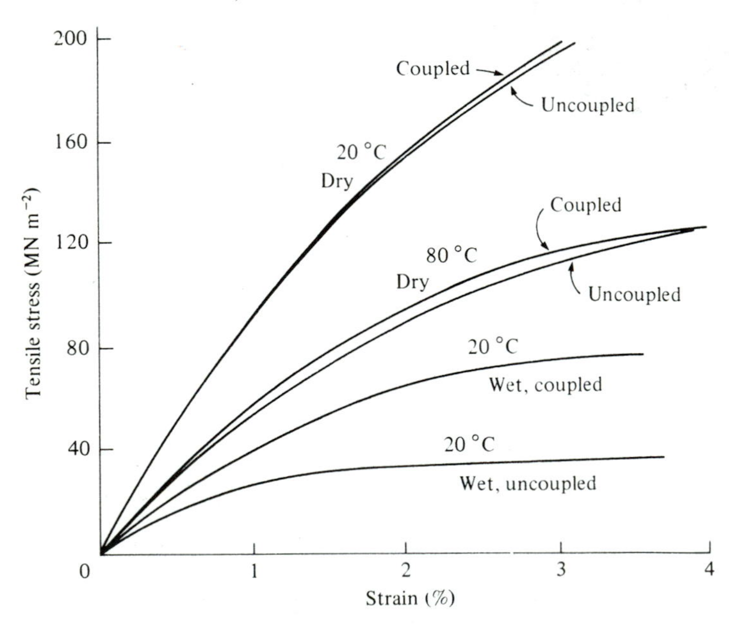

melt fills the mould cavity as illustrated, although it should be noted that the actual flow fields within the melt behind the expanding melt front are more complex and lead to through-thickness variations in fibre orientation (see Fig. 4.19). The changes in fibre orientation lead to different properties within the material. An example is given later in this section.

Typical tensile stress–strain curves for an injection moulded glass-filled nylon 6.6 ($V_f = 0.17$) are shown in Fig. 9.16. They illustrate some important features of the properties of this group of materials. The tests were made on dumbell-shaped specimens which were moulded directly into final shape with an end-gated mould so that melt filling occurred from one end and the melt flow was channelled along the length of the mould. This results in some preferred fibre orientations parallel to the axis of the test specimen. Three effects are evident in Fig. 9.16. (i) The elastic modulus and strength is strongly dependent on temperature. This is due to the temperature dependence of the matrix properties. (ii) The absorption of water results in a very large reduction in the modulus and strength at 20 °C. This can be attributed to changes in the matrix and at the fibre–matrix interface (see Section 10.4). (iii) Fibre–matrix coupling has a relatively small effect on the strength and modulus of dry materials but a major effect on wet materials. The last

Fig. 9.17. Creep curves for glass-filled nylon ($V_f = 0.25$) at 23 °C in air. (From Darlington & Smith 1978.)

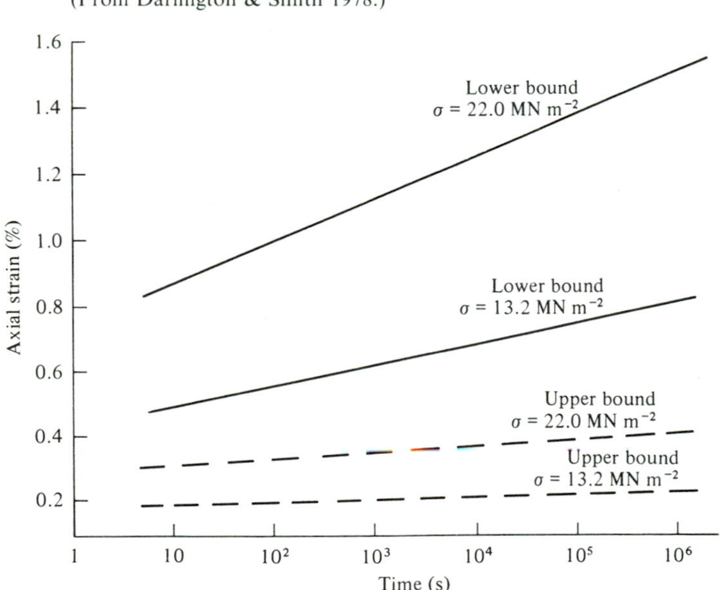

effect illustrates the importance of a coupling agent in providing resistance to deterioration in properties in moist environments. Although all polymers absorb moisture the moisture effect is particularly pronounced in nylon-based materials.

The large non-linear deformations evident in the stress–strain curves in Fig. 9.16 for tests on dry nylon-based materials is due to creep and deformation of the matrix. This is in contrast to polyester–glass CSM where a large part of the non-linear displacement is attributable to resin and interface cracking. Thermoplastics can undergo large amounts of plastic flow before fracture (see Table 2.6) and composite materials with thermoplastic matrices are particularly subject to creep. The effect is most pronounced in low V_f materials and in tests normal to the predominant fibre orientation. Some typical creep curves for glass-filled nylon ($V_f = 0.25$) are shown in Fig. 9.17. The test samples were cut out of a flash-gated square plaque parallel and perpendicular to the injection direction. The mould geometry is similar to that illustrated in Fig. 9.15 except that the gate extends across the full width of the edge of the plaque so that the melt front moves forward as a straight line parallel to the gate. The tensile creep curves were measured at two stress levels. The upper and lower bounds in Fig. 9.17 correspond to specimens tested parallel and perpendicular to the injection direction respectively although the upper bound data were actually obtained from injection moulded dumbell-shaped specimens. The creep rate is clearly very dependent on injection direction and also on the applied stress. The effect of temperature on the stiffness or creep modulus for upper and

Fig. 9.18. Temperature dependence of 100 s tensile creep modulus at 0.10% 100-second creep strain. (From Darlington & Smith 1978.)

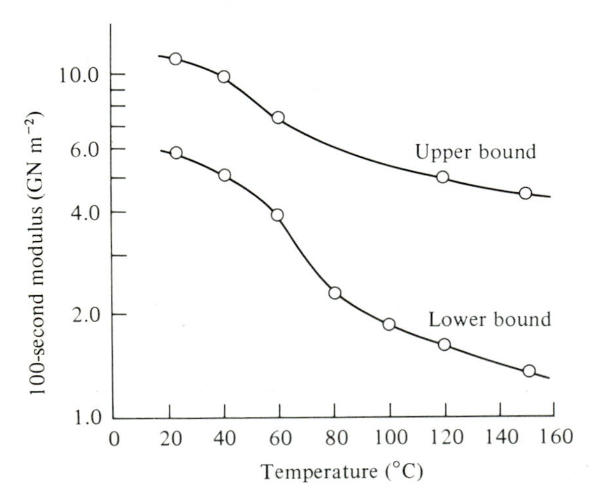

lower bound specimens is shown in Fig. 9.18. The 100-second creep modulus is obtained by measuring the stress required to produce a strain of 0.10% after a time of 100 s has elapsed. Although somewhat arbitrary it is a useful design parameter and provides a direct and accurate method of representing the effect of material variables such as fibre orientation and test variables such as temperature.

The changes in fibre orientation produce corresponding changes in the strength properties which can be understood in terms of the approach used in Sections 9.2 and 9.3. The variations in properties within a moulding lead to considerable complications in designing products to be made by injection moulding. However, they also offer the possibility of achieving optimum product performance by controlling the geometrical and moulding parameters so that the fibres are oriented throughout the moulding to provide maximum load bearing capability.

References and further reading

Berry, J. P. & Stanford, J. L. (1977) Time–temperature moisture effects on the mechanical properties of glass reinforced nylon 6.6. *Reinforced Thermoplastic II*, Plastics and Rubber Institute Conference, paper 11.

Bowyer, W. H. & Bader, M. (1972) On the reinforcement of thermoplastics by imperfectly aligned discontinuous fibres. *J. Mater. Sci.* **7**, 1315–21.

Chen, P. E. (1971) Strength properties of discontinuous fiber composites. *Polym. Engng Sci.* **11**, 51–6.

Darlington, M. W. & Smith, G. R. (1978) Mechanical properties of glass fibre and mineral reinforced polyamides 6 and 6.6. *Plast. Rubber: Mater. Applics*, **3**, 97–108.

Dingle, L. E. (1974) Aligned discontinuous carbon fibre composites. *Proceedings of the Fourth International Conference on Carbon Fibres, their Composites and Applications*, paper 11. Plastics Institute, London.

Halpin J. C. & Kardos, J. L. (1978) Strength of discontinuous reinforced composites. *Polym. Engng. Sci.* **18**, 496–504.

Hancock, P. & Cuthbertson, R. C. (1970) Effect of fibre length and interfacial bond in glass fibre–epoxy resin composites. *J. Mater. Sci.* **5**, 762–8.

Johnson, A. F. (1979) *Engineering Design Properties of GRP*, British Plastics Federation, London.

Kelly, A. (1970) Interface effects and the work of fracture of a fibrous composite. *Proc. R. Soc. Lond.* A. **319**, 95–116.

Lees, J. K. (1969) A study of the tensile strength of short fiber reinforced plastics. *Polym. Engng Sci.* **8**, 195–201.

Owen, M. J. (1970) Fatigue testing of fibre reinforced plastics. *Composites* **1**, 346–55.

Owen, M. J. & Found, M. S. (1972) Static and fatigue failure of glass fibre reinforced polyester resins under complex stress conditions. *Faraday Special Discussions of the Chemical Society*, no. 2. 77–89.

Ramsteiner, F. & Theysohn, R. (1979) Tensile and impact strengths of unidirectional, short fibre-reinforced thermoplastics. *Composites* **10**, 111–19.

10 Other topics

10.1 Introduction

In this chapter four scientifically and technologically important aspects of composite materials are discussed in outline. It is not possible in the context of this book to cover them in any detail because each really requires a lengthy treatment. Brief reference is made to some of the underlying concepts and a small number of references is given to provide an introduction to the abundant technical literature on these subjects. Reference is made also to earlier chapters to give an indication of how the more complicated topics dealt with in this chapter can be understood in terms of more simple concepts. This chapter is not comprehensive and the topics chosen are only examples from a wider range of problems associated with the design, manufacture and application of composite materials and products.

Most of the topics are related to the use of composite materials in more demanding loading and environmental conditions than those considered previously. Thus, the first topic is concerned with the resistance of composite materials to reversible-loading, i.e. fatigue, conditions. Fatigue damage is related to the micromechanisms of deformation and fracture in the material. Two factors are particularly relevant. Firstly, the general degradation of the microstructure which leads to progressive reduction in properties and, secondly, the rate of growth of cracks under fatigue loading. The second topic relates to the response of materials and components to the presence of stress raisers such as holes and notches. Again this depends on the deformation and fracture processes which occur particularly in the regions of high local stress and strain. The ability of the material to 'spread the load' in regions of high stress magnification is a major factor determining its fracture toughness and is directly relevant to both fatigue crack growth and the notch sensitivity of materials.

The third topic deals with the change of properties which occurs in the presence of liquid environments with and without a superimposed external load. The effect of these environments on the fibre, matrix and interface is discussed and a brief reference is made to the creep of composite materials. The final topic is related to the design of new and superior composite materials by combining two or more different types of fibre. These '*hybrid*' composite materials offer the possibility of achieving unique properties and of designing structures which make the most economic use of high performance expensive fibres.

10.2 Fatigue

Fatigue may be defined as the ultimate failure of a material or component by the application of a varying load whose maximum amplitude, if continuously applied, is insufficient to cause failure. One of the common ways of representing fatigue is the use of the *S–N* curve illustrated in Fig. 10.1. An alternating stress is applied to the material and the number of cycles to failure (N) is determined as a function of the stress amplitude (S). The slope of the *S–N* curve is a measure of the resistance of the material to fatigue and the actual shape varies from one material to another.

In fundamental terms fatigue is due to the irreversible processes which occur when a cyclic load is applied to a material. The extent of fatigue damage and its importance is dependent primarily on the stress level at which irreversible damage occurs relative to the stress for complete failure. Consider two materials with stress–strain curves shown in Fig. 10.2. One material is almost perfectly elastic to fracture whereas the other undergoes plastic or viscoelastic flow at about $0.3\,\sigma^*$. The elastic

Fig. 10.1. Schematic representation of *S–N* curves. (N is the number of cycles to failure.) Note that alternating stresses may be tensile, compressive or shear.

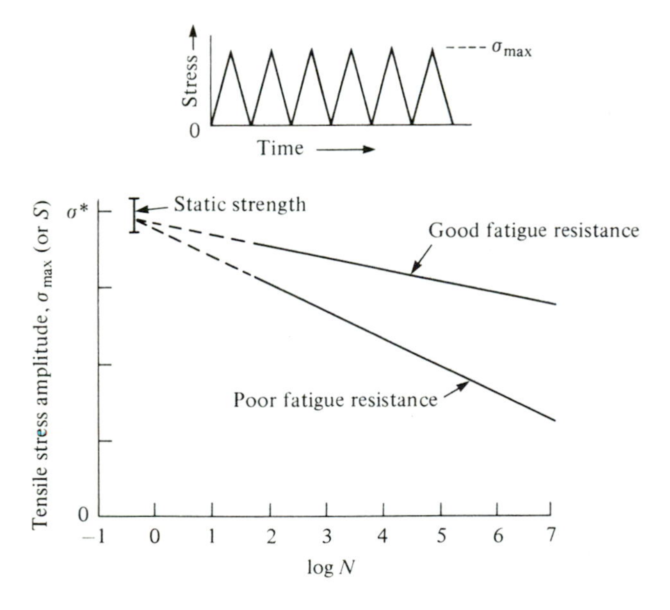

material will be insensitive to alternating stresses below σ^* and the *S–N* curve will be almost flat, i.e. the fatigue resistance is very good. The other material will start to deform at relatively low stresses and fatigue damage will develop continuously. The amount of damage will increase with increasing load and the material will have a poor resistance to fatigue.

When this simple principle is applied to composite materials it is clear that the response depends on the fibre arrangement and volume fraction as well as on the matrix and fibre properties. All these factors determine the way that the load is distributed between the fibre, matrix and

Fig. 10.2. Stress–strain curves. *A* is almost perfectly elastic to fracture; *B* undergoes plastic or viscoelastic flow.

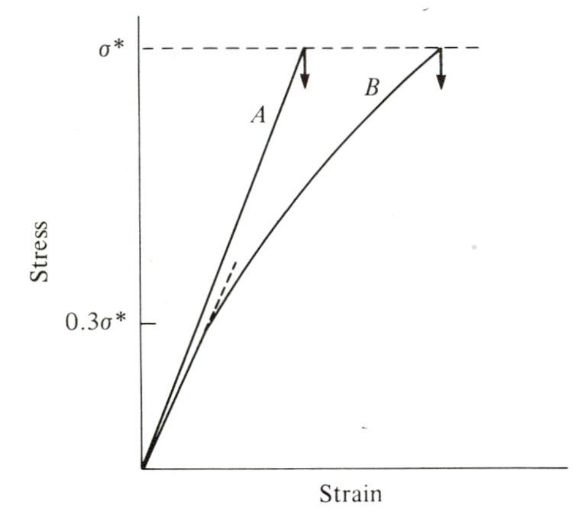

Fig. 10.3. *S–N* curve of unidirectional laminae, Type I carbon fibres in epoxy and polyester resins ($V_f \approx 0.4$), tested in repeated tension in longitudinal direction. (From Beaumont & Harris 1971.)

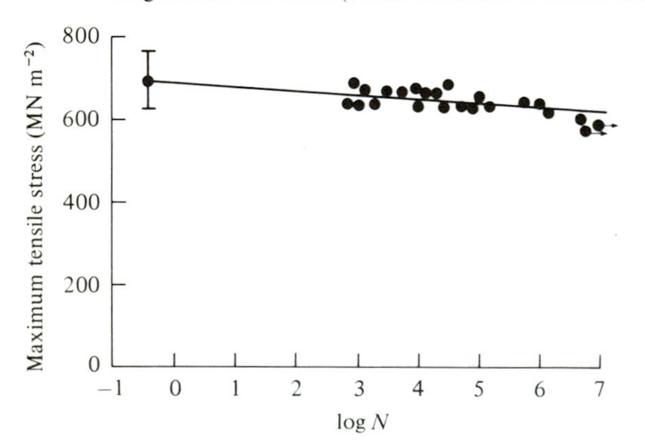

fibre–matrix interface. As an example compare the response to fatigue loading of a unidirectional high V_f carbon fibre–epoxy composite material tested parallel to the fibres with the response of a random in-plane glass fibre–polyester resin material. Typical *S–N* curves are shown in Figs. 10.3 and 10.4. In the aligned carbon fibre material (Fig. 10.3) the load is carried almost entirely by the fibre and little irreversible damage occurs until fibre fracture is initiated. Subsequent unloading and loading cycles result in a small redistribution of the load in the region of the broken fibres and some fatigue damage may develop. This will occur only at stresses close to ultimate fracture so that the fatigue resistance is very good. In the random glass fibre material (Fig. 10.4) microcracking by debonding occurs in transversely oriented fibre bundles at relatively low stresses (see Fig. 9.10) and this marks the onset of clearly defined irreversible damage. Cracking may be preceded by resin flow. During fatigue cycling the transverse cracks propagate and eventually lead to resin cracking which can occur at stresses well below those observed in monotonic loading. The results are well illustrated in Fig. 10.4 which shows that fatigue loading results in a reduction in the stresses at the onset of debonding and resin cracking as well as in the final fracture strength.

Fig. 10.4. *S–N* curves of chopped-strand mat–polyester resin tested in fully reversed loading showing the various stages of failure. (From Smith & Owen 1968.)

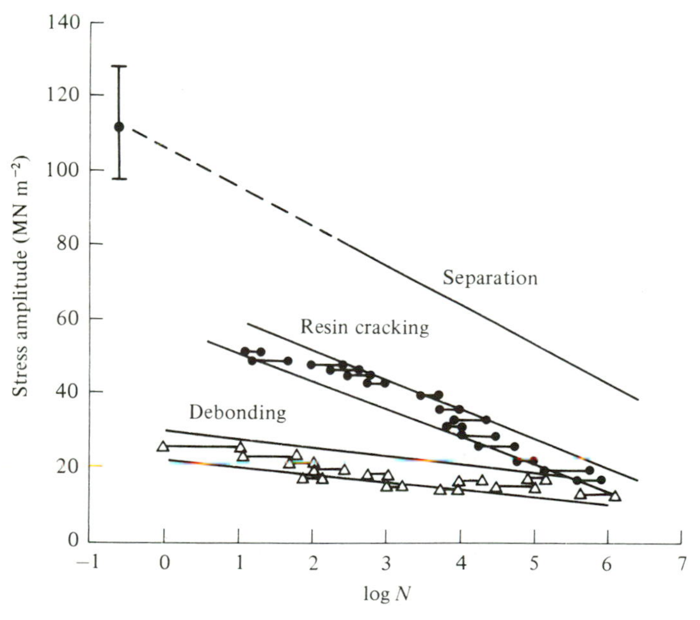

Although the nature of the irreversible processes depends on the composite material and the loading conditions, the principle of progressive fatigue damage can be understood by reference to the simple example in Fig. 10.5. The diagram represents part of a fibre bundle or lamina oriented with the fibre axis normal to the applied stress. The bundle or lamina is constrained by adjacent bundles so that the growth of a crack under monotonic loading occurs only under increasing strain conditions. Suppose that the bundle is loaded by a simple cyclic stress as illustrated in Fig. 10.1. In the first half cycle the deformation and fracture processes which occur depend on the stress amplitude. Reference to Section 7.4 on transverse tensile strength indicates that the initial response is linear (Fig. 7.19*b*). As the stress increases, non-linear effects arise owing to the viscoelastic properties of the resin and, at a later stage, to debonding and resin cracking. Thus, at very low stress amplitudes, when the response is fully elastic, fatigue damage will not develop.

Fig. 10.5. Schematic representation of the nucleation and growth of fatigue cracks in a transversely oriented bundle or lamina surrounded by differently oriented bundles or laminae.

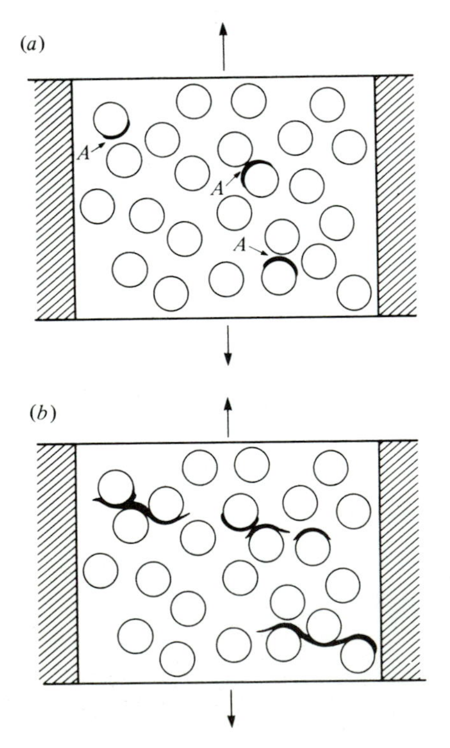

When the stress amplitude is increased viscoelastic flow occurs preferentially between closely spaced fibres (at *A* in Fig. 10.5*a*) because of the high strain magnification in these regions. This flow will not be fully reversible when the stress is reduced and during cyclic loading additional stress and strain concentrations develop which lead to the initiation of debonding at applied stresses below those observed in monotonic loading (Fig. 10.5*a*). The debonding cracks grow during cyclic loading because some flow occurs at the crack tip during the loading half of the cycle which is not fully reversed on unloading (Fig. 10.5*b*). As in uniaxial tensile tests the cracks nucleate and propagate in regions of closely spaced fibres by the growth and coalescence of individual fibre debonds. When the fibres are widely spaced the growth of the crack from one fibre to the next depends on the fatigue crack growth resistance of the matrix.

It follows from this simplified model that the fatigue properties will depend on the temperature of the test and the cyclic loading frequency since both these factors affect the amount of matrix flow which occurs. An important additional effect is viscous dissipative heating during cyclic loading leading to a rise in temperature which can be between 25 °C and 50 °C at high frequencies. The magnitude of the rise depends on specimen geometry and the efficiency of heat transfer to the surroundings. Thus, carbon fibre composite materials show lower temperature rises because of the high thermal conductivity of the fibre.

Similar reasoning to the above can be used to explain the fatigue behaviour of more complicated fibre arrangements. Progressive damage may involve intralaminar and interlaminar processes and the rate of damage build-up depends on the effective stresses causing different forms of microdamage.

Fig. 10.6. Macroscopic appearance of fatigue damage after resin cracking in a chopped-strand mat laminate. (From Owen 1970.)

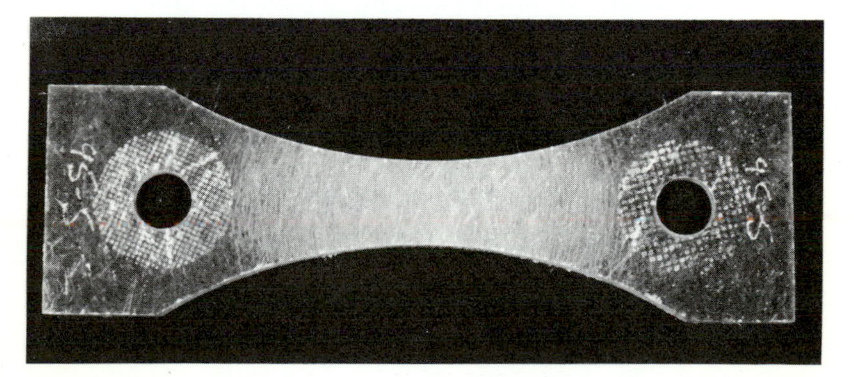

A common feature of the fatigue failure of composite materials, particularly when the test specimens are uniformly stressed, is the occurrence of damage over a large volume of the material as illustrated in Fig. 10.6. More severe localised damage occurs at a later stage of the 'fatigue life' and results in final separation. The large scale damage leads to three important effects which can be summarised as follows: (i) the modulus of elasticity decreases progressively during the fatigue life, (ii) the hysteresis or damage loop during cyclic stressing becomes progressively more pronounced, and (iii) the residual strength of the material decreases progressively with the number of cycles. It will be necessary to develop cumulative damage rules to predict the long term performance of components subject to fatigue damage.

Finally, it should be noted that the failure of engineering components and structures subject to fatigue loading often develops from local regions of high stress or strain associated with, for example, sharp changes in cross-section, holes and accidental damage. The micro-mechanisms of failure in these regions will be the same as those described in earlier chapters and they result in the slow growth of cracks as described in the next section. Again crack growth laws are required to relate the crack length to the number of loading cycles.

10.3 Notch sensitivity and fracture energy

Brief reference has been made elsewhere to the effect of holes and notches in producing local regions with modified stress fields. Consider the simple example of a flat plate of isotropic elastic material with a circular hole (Fig. 10.7*a*) subjected to a uniform tensile load P. The average stress $\sigma = P/A$, where A is the cross-sectional area of the plate, and the stress field in the vicinity of the hole is non-uniform. The maximum uniaxial tensile stress occurs at the edges of the hole at M and N and is 3σ. In other words the stress concentration factor is 3. The stress concentration factor varies with the shape of the hole. For an elliptical hole oriented with the major axis normal to the applied stress (Fig. 10.7*b*) it is given by

$$K = \sigma_{\max}/\sigma = 1+2c/b \tag{10.1}$$

or

$$K = 1+2(c/\rho)^{\frac{1}{2}} \tag{10.2}$$

where ρ is the radius of curvature at the tip of the ellipse. Thus, the stress concentration factor will be very large for long flat elliptical holes or cracks. Although equation (10.1) and (10.2) apply only to isotropic materials the general principles are relevant to all materials.

The main ideas relating to the notch sensitivity of materials are illustrated schematically in Fig. 10.8. A flat plate containing two edge notches is tested in uniaxial tension. The fracture strength of such a plate is normally defined as P^*/Wt where P^* is the load at fracture, W is the width, and t the thickness, of the plate. The presence of the notches leads to a reduction of strength for two reasons. Firstly, the effective cross-sectional area of the plate is reduced to $(W-2c)t$ and secondly, stress concentrations are produced at the notch tips as explained above. These two effects are distinguished in Fig. 10.8*b* by plotting the net cross-sectional area 'strength' $\sigma_n^* = P^*/(W-2c)t$ against c. For materials which are insensitive to the stress concentration effects of the notch (notch insensitive) σ_n^* is independent of notch depth. In contrast,

Fig. 10.7. Stress concentrations around holes in a plate subjected to a uniform tensile load.

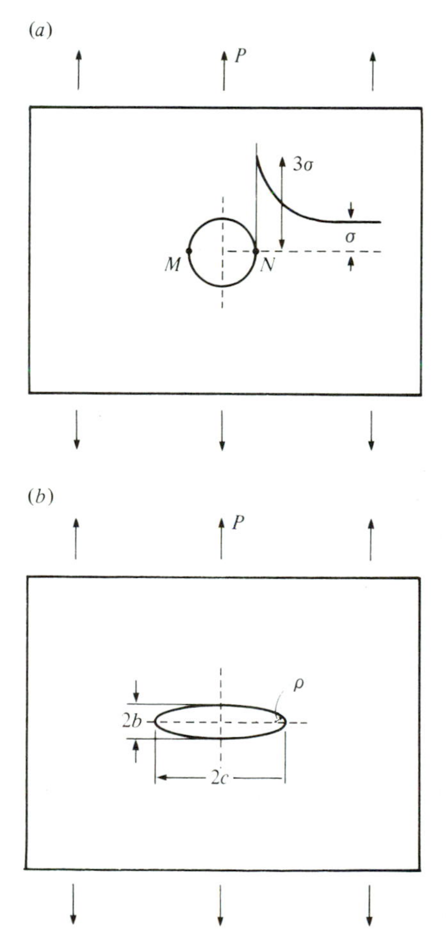

the strength of notch sensitive materials is strongly dependent on notch depth and decreases rapidly as the depth increases. The rate of decrease is greater for sharp notches than blunt notches (see equation (10.2)).

Notch sensitivity is related to the response of the material at the root of the notch to the high local stresses. If plastic flow and microcracking processes occur without the notch depth increasing, the stress concentration is reduced because the load is distributed over a larger volume of material. On the other hand, if no deformation occurs the stress concentration effect dominates and crack growth takes place at the root of the notch. The effects can be understood in terms of the way that the tip of the notch is blunted by deformation and microcracking so reducing the stress concentration factor.

The response of composite materials to the presence of notches depends on the fibre geometry and material properties. Some examples are illustrated schematically in Fig. 10.9. In a unidirectional lamina (Fig.

Fig. 10.8. Strength properties of notch sensitive and notch insensitive materials.

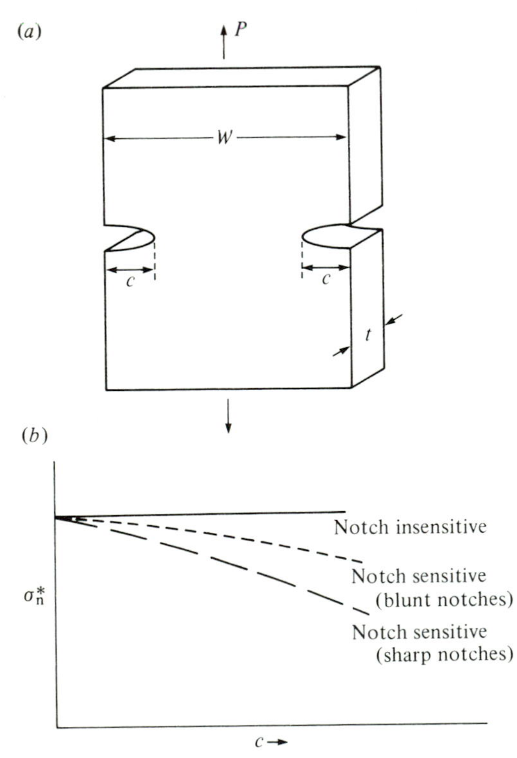

10.9*a*) in which the fibres are weakly bonded to the matrix the stress concentration effects of the notch lead to fibre debonding at stresses well below those required to cause fibre fracture (see Section 7.2). The debonding spreads along the fibres and the resulting elastic relaxation due to unloading of the fibres causes the notch to open up and become blunt so removing the stress concentration (Fig. 10.9*b*). The material is, therefore, almost completely notch insensitive. If the fibre is strongly bonded to the matrix, stress relaxation can occur only by resin shear or fibre fracture. Thus, the notch sensitivity increases with increasing bond strength.

Fig. 10.9. Effect of fibre geometry on notch sensitivity. (*a*) Unidirectional lamina tested in longitudinal tension. (*b*) as for (*a*) showing cracks parallel to fibres and blunting of notch (notch insensitive). (*c*) Unidirectional lamina tested in transverse tension (notch sensitive). (*d*) CSM tested in uniaxial tension illustrating formation of damage zone owing to debonding and matrix cracking at root of notch. (*e*) as for (*d*), damage zone has grown and a crack has formed at the root of notch. (*f*) $\pm 45°$ angle-ply laminate tested in uniaxial tension showing damage zones due to cracking parallel to fibres.

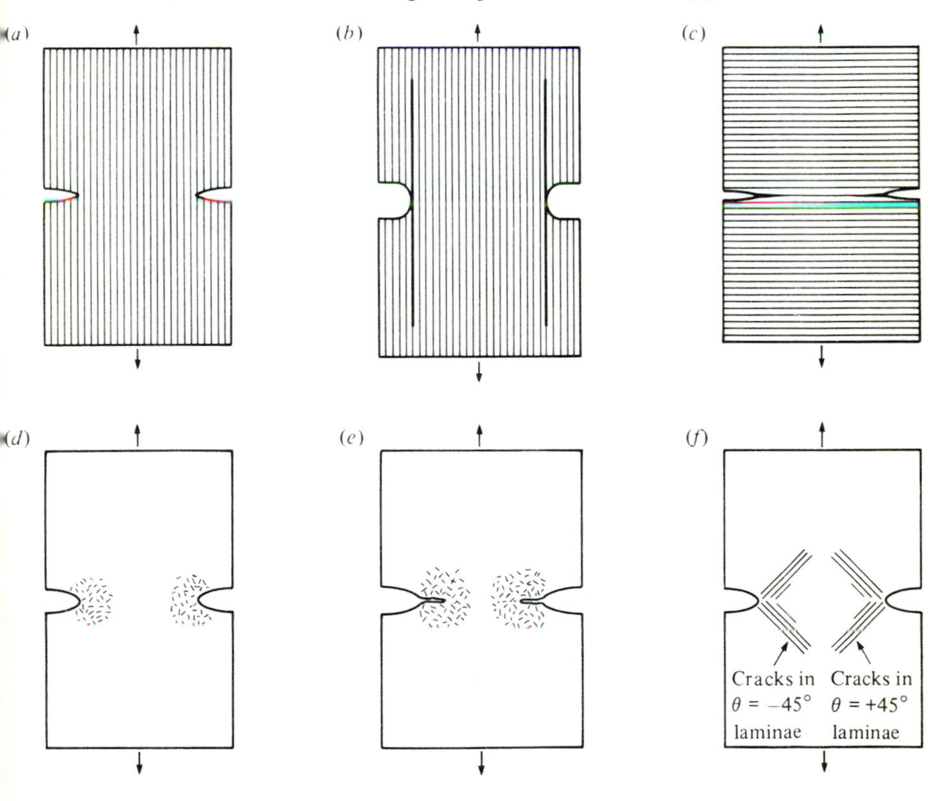

In a unidirectional lamina tested normal to the fibres, (Fig. 10.9*c*) there are no micromechanisms which produce effective crack blunting and the material is very notch sensitive in this direction. The small amount of resin flow at the crack tip has a negligible effect, particularly at high V_f.

It will be evident from Section 9.3 that in a random fibre material, such as chopped-strand mat–polyester resin laminates, there are many micromechanisms which lead to macroscopic displacements and hence result in relaxation of the load at the tip of a notch. These include debonding, resin shear, matrix and fibre fracture and fibre pull-out. When a notched sample is tested a zone of damage develops at the root of the notch and gradually spreads as the load is increased (Figs 10.9*d* and 10.9*e*). If the zones spread across the width of the test specimen before final fracture occurs, the strength of the material will be relatively insensitive to the notches. In wide specimens and in large structures, however, the damage zones are constrained within the material. The strains at the notch are then sufficient to cause complete fracture of the material at the root and hence a crack grows from the notch. Thus, the material will show a degree of notch sensitivity depending on notch depth and specimen width.

A similar sequence of events occurs in angle-ply laminates (Fig. 10.9*f*) except that the micromechanisms of failure are dominated by the fibre

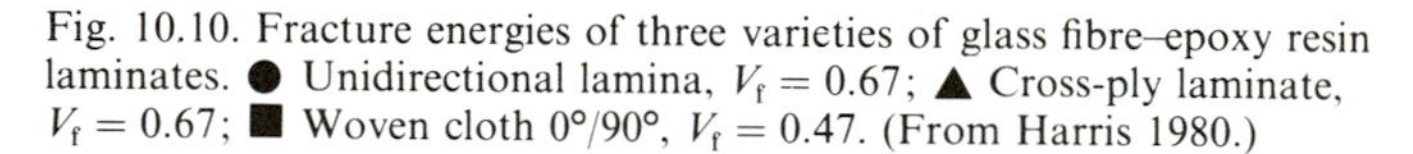
Fig. 10.10. Fracture energies of three varieties of glass fibre–epoxy resin laminates. ● Unidirectional lamina, $V_f = 0.67$; ▲ Cross-ply laminate, $V_f = 0.67$; ■ Woven cloth 0°/90°, $V_f = 0.47$. (From Harris 1980.)

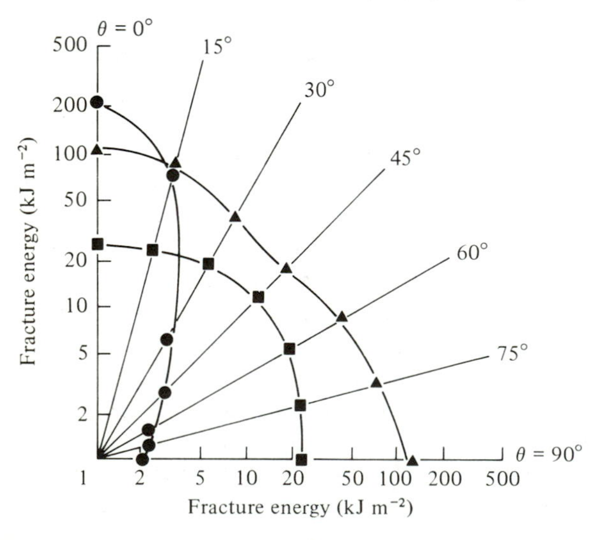

geometry. As described in Sections 8.3 and 8.4, large displacements result from transverse and intralaminar shear cracking. The amount of cracking is controlled by the interlaminar shear strength. The extent of the damage zone and hence the notch sensitivity depends on notch geometry and material properties and, in particular, on the ply-angle and stacking sequence.

The notch sensitivity of materials is closely related to their resistance to fracture. For design purposes it is necessary to predict the overall loads which can be applied to a structure without catastrophic propagation of pre-existing flaws or flaws generated during service. This requirement is the primary reason for the development of fracture mechanics; a subject which is beyond the scope of this book. In fracture mechanics the most important material parameter is the fracture energy or the amount of work done when a crack propagates. Since all the microdeformation and microcracking processes which precede final separation contribute to the work of fracture, it follows that the fracture energy varies widely from one composite material to another. This is amply illustrated in Fig. 10.10 for tests on three different glass fibre–epoxy resin materials. In these tests the fracture energy was measured using a Charpy impact test. This involves breaking a notched bar test sample into two pieces using a pendulum impact head and measuring the work done. Thus, some materials have a very large fracture energy and others very small. As expected the greatest variation in fracture energy with orientation is obtained in unidirectionally aligned materials.

10.4 Deterioration of properties owing to environmental conditions

So far it has been assumed that the properties of composite materials are independent of the thermal and chemical environment although some reference has been made to the effect of temperature. There are three important factors related to temperature. Firstly, differential expansion between fibre and resin and between differently oriented laminae in a laminate can result in internal stresses as described briefly in Section 5.6. Secondly, changes in temperature result in changes in the properties of the constituent materials, in particular the matrix. In the absence of a superimposed load the effect of temperature on the matrix properties is usually reversible unless the temperature approaches the glass transition of crystallisation temperature. Changes in matrix properties affect the short term strength and modulus of those composite material properties, such as longitudinal compressive modulus of unidirectional laminae, which are strongly dependent on the matrix. Thirdly, the resistance of materials to strain under load (creep resistance) is strongly dependent on temperature because of the viscoelastic

Table 10.1. *Aspects of property deterioration in polymer-based composite materials*

	Reversible changes	Irreversible changes
Resin	Water swelling Temperature flexibilising Physical ordering of local molecular regions	(1) Chemical break-down by hydrolysis (2) Chemical break-down by UV radiation (3) Chemical break-down by thermal activation (4) Chemical breakdown by stress induced effects associated with swelling and applied stress (5) Physical ordering of local molecular regions (6) Chemical composition changes by leaching (7) Precipitation and swelling phenomena to produce voiding and cracks (8) Non-uniform de-swelling to produce surface cracks and crazes (9) Chemical effect of thermoplastic polymer content on long term stability
Interface	Flexibilising interface	(1) Chemical break-down as above 1, 2, 3, 4 (2) Debonding due to internally generated stresses associated with shrinkage and swelling and the applied stress (3) Leaching of interface
Fibre	—	(1) Loss of strength due to corrosion (2) Leaching of fibre (3) Chemical break-down by UV radiation

Note: The relative importance of these effects will depend on environmental test conditions.

properties of the matrix. Flow and irreversible microcracking can occur over prolonged periods which may eventually lead to complete separation. Again those properties which depend on the matrix are most affected and the problem is of particular concern in short fibre reinforced thermoplastics.

A gaseous or liquid environment may produce profound changes in properties of composite materials. These can best be understood in terms of the effect of these environments on the individual constituents, i.e. fibre, matrix and interface. A summary of the main changes which can occur is given in Table 10.1. Although the list has been divided into reversible and irreversible changes, it is unlikely that any of these processes is truly reversible.

The changes in matrix properties are due to the diffusion of the environment through the polymer. The stability of the polymer depends on the chemistry and conformation of the molecules, and in thermosetting polymers is strongly dependent on the cross-linked structure. As an example consider the diffusion of moisture into a carbon fibre–epoxy resin laminate from a humid environment. For a given resin the equilibrium amount of water absorbed depends on the relative humidity of the environment and the temperature. The rate of uptake of water also depends on these factors and on the diffusion coefficient of the resin. The presence of the moisture results in swelling of the resin which is partly constrained by the presence of the fibre. The degree of constraint depends on the fibre geometry and volume fraction. Under non-equilibrium conditions the water content is not uniform and the concentration gradients lead to stress and strain gradients in the material. Similarly, even when the moisture content is uniform, the different swelling characteristics of adjacent laminae in a laminate leads to a build-up of internal stresses.

Moisture has the effect of flexibilising the epoxy resin or, in other words, reducing the elastic moduli. As mentioned in Chapter 7 changes in matrix moduli have little effect on longitudinal tensile strength and modulus because the matrix plays only a minor role. However, the effect on the longitudinal compressive strength and the interlaminar and intralaminar shear properties may be substantial because the resin contribution is large. This is particularly true at elevated temperatures since the presence of the moisture also reduces the glass transition temperature of the resin. Thus, an uptake of 1.5% water produces a small reduction in the interlaminar shear strength of carbon fibre epoxy resin laminates at room temperature but the reduction can be 50–60% at temperatures between 100 °C and 130 °C.

The decrease in elastic modulus of epoxy resin due to uptake of water is reversible in that the modulus returns to its original value when the environmental conditions change and the moisture diffuses out. However, there are many processes associated with moisture absorption, as listed in Table 10.1, which lead to irreversible changes in property. These become more significant after prolonged exposure and are strongly dependent on resin chemistry and, for thermosetting resins, on the degree of cure. There is direct experimental evidence for micromechanical break-down associated with resin and interface cracking. Many elegant model experiments have been reported by Ashbee and his colleagues (see e.g. Ashbee & Wyatt 1969). The irreversible changes can lead to significant degradation of physical and mechanical properties.

The diffusion of moisture through the resin matrix is affected by the fibres in a number of ways. Firstly, the geometry of the diffusion path is dependent on the fibre arrangement and volume fraction. Secondly, the interface has different diffusion characteristics and a wicking or capillary action can occur along the interface causing rapid diffusion. Thirdly, the irreversible process mentioned above lead to the nucleation and growth of microcracks at the interface and in the resin which then provide an easy path for the environment. The diffusion processes and hence the rate of deterioration are increased in the presence of applied tensile stresses. The exact mechanisms are not understood but it seems likely that hydrostatic tensile stresses will aid swelling processes in the resin and hence increase the diffusion rates and total water uptake. Stresses will undoubtedly have an effect when cracking processes are involved.

In addition to fine scale microcracking processes, the ingress of moisture can lead to the growth and propagation of large cracks. A particularly good example is the blistering of some polyester laminates which with some resin–fibre coating combinations can occur at temperatures as low as 35 °C after prolonged exposure to water. The cracking is due to an osmotic pressure effect. The water diffuses through the resin and separates out at voids. In the process it dissolves any soluble materials in the surrounding glass and polyester. The resin acts as a semi-permeable membrane and the osmotic pressure develops inside the hole because the water continues to diffuse to the regions with the largest concentration of soluble material. The pressures are eventually sufficient to cause rupture and blister formation.

The environment may also affect the strength properties of the fibres. Thus, for example, glass fibre is susceptible to water leaching of the soluble oxides such as K_2O and Na_2O. After prolonged exposure, leaching results in surface pits and reduction of properties. The presence of dissolved salts in the water around the fibre can lead to the generation of osmotic pressures which may eventually produce debonding at the interface even in the absence of an applied stress. Coupling agents reduce the rate of water attack on the glass surface.

Additional effects occur in the matrix and at the fibre surface when an acid environment is used. There is a pronounced 'stress-corrosion' effect in glass fibre composite materials when tests are carried out under the combined influence of stress and an acid environment. The effect is usually attributed to a hydrogen ion exchange process of the form

$$\text{—Si—O—Na} + H^+ \rightarrow \text{—Si—O—H} + Na^+$$

This results in weakening of the glass by the creation of large tensile stresses at the surface as a result of shrinkage of the surface layers.

10.5 Hybrid composite materials

Composite materials containing two or more different types of fibre, for example glass and Type I carbon or glass and Kevlar, are called 'hybrid' composite materials. They can be made using the manufacturing methods summarised in Table 1.5 although there are some limitations in achieving fibre–matrix compatibility with some of these methods since two different fibres are involved. Numerous fibre arrangements can be envisaged but systematic work on physical and mechanical properties has been limited to a few simple arrangements. Two examples are illustrated in Fig. 10.11. In the unidirectionally aligned material sketched in Fig. 10.11*a* the individual fibres or fibre bundles are intermingled in a random manner. Alternatively, the intermingling could be organised in an ordered array. The matrix is common to both

Fig. 10.11. Hybrid composite materials (*a*) unidirectional lamina with intermingled fibres or fibre bundles (*b*) unidirectional laminate with alternate homogeneous laminae.

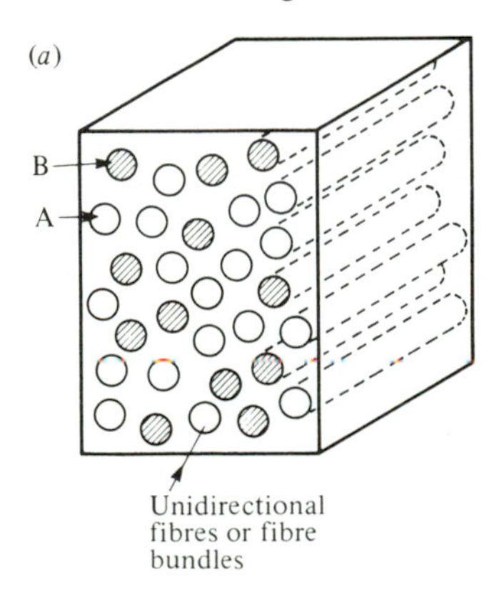

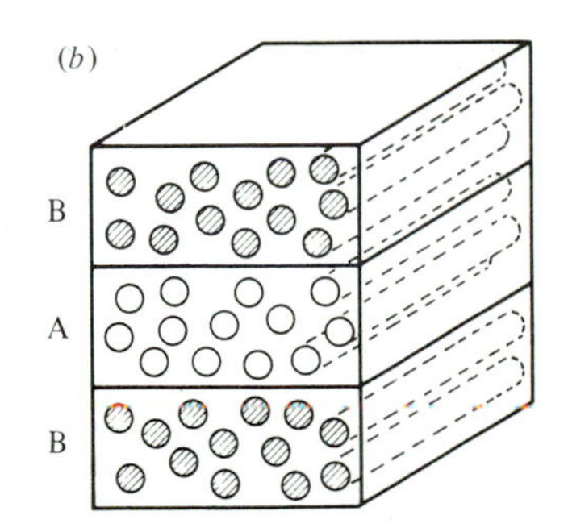

sets of fibres. The laminate in Fig. 10.11*b* consists of alternate laminae of the two fibres, each lamina being homogeneous. All the fibres are parallel to each other. Many variants of this simple arrangement can be envisaged such as cross-ply and angle-ply configurations. The matrix in each lamina may be different although compatibility between matrices is essential if good interlaminar properties are required.

The importance of hybrid composite materials is related to (i) the cost savings which can be achieved by replacing expensive carbon fibres by less expensive glass fibres (ii) the wider range of physical and mechanical properties which can be obtained by optimising the choice of the fibres used and their volume fractions, and (iii) the possibility of obtaining unique properties, singly or in combination which are not readily obtained when using one type of fibre alone.

Analytical methods for predicting the elastic properties of hybrid materials are being developed although it will be evident from Chapter 6 that this is likely to be a complex problem. A few simple points are worth noting. In laminates made up of unidirectional laminae the elastic properties can be predicted by using the theory outlined in Chapter 6. Thus, the properties of each lamina are determined and the laminate properties are calculated by summation as described in Section 6.3 for cross-ply and angle-ply laminates. The same considerations about symmetric lay-up sequences apply if out-of-plane coupling forces are to be avoided. There is the additional constraint that the different fibre types have to be considered as well as fibre angles and the thicknesses

Fig. 10.12. Effect of stacking sequence on bending stiffness of hybrid materials.

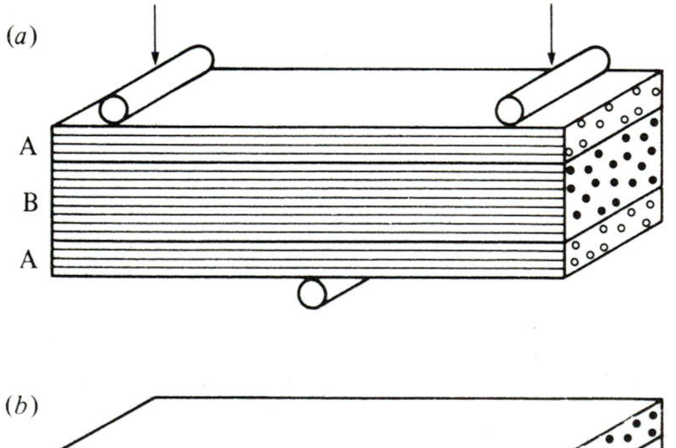

of the laminae. For in-plane loading the elastic properties will not be significantly affected by the position of the different fibre layers provided the laminates remain symmetric and the volume fractions are constant. However, for out-of-plane loading this is not so as will be evident from the simple example illustrated in Fig. 10.12. The laminate with the stiffer fibres ($E_A > E_B$) in the outer layers (Fig. 10.12*a*) has a greater bending stiffness, in the direction indicated, than the laminate with the stiffer fibres in the inner layers (Fig. 10.12*b*). This is because the outer layers take a greater proportion of the applied load. The stiffening effect is used in practice by applying skins of carbon fibre material to a core of glass fibre material in beams and struts. The principle is the same as that used in the design of I-beams.

Except for one or two geometrical arrangements, the elastic properties of intermingled fibre hybrid materials are not amenable to theoretical analysis at the present time particularly when the fibres are randomly oriented. For the simple example illustrated in Fig. 10.11*a* the 'rule of mixtures' provides a good approximation of the longitudinal stiffness since the longitudinal strain in the two different fibres and the resin is the same (see Section 5.2). However, in general, the elastic properties can be determined only by experiment. The main ideas outlined in previous chapters will apply and, as in the case of single fibre composite material, the matrix will have a dominant effect on many of the properties.

The strength properties of hybrid materials are little understood also. This is not surprising in view of the difficulty of predicting the properties of single fibre materials. Two points are of particular interest. Firstly, the primary effect of hybridisation on the longitudinal strength of aligned fibre laminates is related to the strains to failure of the different fibres. Thus, for the laminate in Fig. 10.11*b* the laminae with the lowest strain to failure will fail first even though they may have a higher strength. The ultimate failure will depend on the way that the load is redistributed after initial failure. Similar arguments apply to those used in Section 7.2 in considering the effect of different fibre and matrix strains. When cracking in one lamina does not lead to complete failure of the laminate multiple cracking will follow, the characteristics of which depend on the interlaminar bond. When first cracking leads to laminate failure the strength is less than the 'rule of mixtures' strength and is determined by the failure strain of the first lamina. Various proposals have been made to account for occasional experimental observations of longitudinal tensile strengths greater than the rule of mixtures prediction. These include the possibility of built-in internal stresses due to differential contraction of the different layers after curing at elevated temperatures.

The second feature of interest is the contribution of different fibres to the fracture resistance and fracture energy. These parameters are related to the patterns of microdeformation and microfracture which occur in the presence of cracks and other stress raisers. It follows that by mixing fibres with different moduli, strains to failure and interface strengths it should be possible to design materials with different energy absorbing capabilities.

References and further reading

Advanced Composite Materials – Environmental Effects (1978) ASTM Special Technical Publication 658, ed. J. R. Vinson. American Society for Testing Materials.

Ashbee, K. H. G. & Wyatt, R. C. (1969) Water damage in glass fibre–resin composites. *Proc. R. Soc. Lond.* A **312**, 553–64.

Beaumont, P. W. R. & Harris, B. (1971) The effect of environment on fatigue and crack propagation in carbon-fibre reinforced epoxy resin. *Proceedings of the International Conference on Carbon Fibres, their Composites and Applications*, paper 49. Plastics Institute, London.

Beaumont, P. W. R. & Phillips, D. C. (1972) The fracture energy of a glass fibre composite. *J. Mater. Sci.* **7**, 682–6.

Caprino, G., Halpin, J. C. & Nicolais, L. (1979) Fracture mechanics of composite materials. *Composites* **10**, 223–7.

Cooper, G. A. & Kelly, A. (1967) Tensile properties of fibre-reinforced metals: fracture mechanics. *J. Mech. Phys. Solids* **15**, 279–97.

Dally, J. W. & Carrillo, D. H. (1969) Fatigue behaviour of glass-fiber fortified thermoplastics. *Polym. Engng Sci.* **9**, 434–44.

Dew-Hughes, D. & Way, J. L. (1973) Fatigue of fibre-reinforced plastics: a review. *Composites* **4**, 167–73.

Fatigue of Fibre Reinforced Plastic Composites (1977) Conference Proceedings. Society of Environmental Engineers, London.

Harris, B. & Bunsell, A. R. (1975) Impact properties of glass fibre/carbon fibre hybrid composites. *Composites* **6**, 197–201.

Harris, B. & Ankara, A. O. (1978) Cracking in composites of glass fibres and resin. *Proc. R. Soc. Lond.* A. **359**, 229–50.

Harris, B. (1980) Micromechanisms of crack extenson in composites. *Metal Sci.* **14**, 351–62.

Hogg, P. J. & Hull, D. (1980) Micromechanisms of crack growth in composite materials under corrosive environments. *Metal Sci.* **14**, 441–9.

Johnson, A. F. (1979) *Engineering Design Properties of GRP*. British Plastics Federation, London.

Judd, N. C. W. (1977) Absorption of water into carbon fibre composites. *Br. Polym. J.* March, 36–40.

Kanninen, M. F., Rybicki, E. F. & Brinson, H. F. (1977) A critical look at current applications of fracture mechanics to the failure of fibre reinforced composites. *Composites* **8**, 17–22.

Mandell, J. F. & Wang, S. S. (1975) The extension of crack tip damage zones in fiber reinforced plastic laminates. *J. Comp. Mater.* **9**, 266–87.

Mandell, J. F. (1979) Fatigue crack growth in fiber reinforced plastics.

Proceedings of the 34th SPI/RP Annual Technology Conference, paper 20-C. Society of the Plastics Industry, New York.

Metcalfe, A. G. and Schmitz, G. K. (1972) Mechanisms of stress corrosion in E glass filaments. *Glass Technol.* **13**, 5–16.

Owen, M. J. (1970) Fatigue testing of fibre reinforced plastics. *Composites* **1**, 346–55.

Owen, M. J. & Morris, S. (1971) Fatigue behaviour of orthogonally cross-plied carbon fibre-reinforced plastics under axial loading. *Proceedings of the First International Conference on Carbon Fibres, their Composites and Applications*, paper 51. Plastics Institute, London.

Owen, M. J. & Bishop, P. T. (1975) Prediction of static and fatigue damage and crack propagation in composite materials. *AGARD Conference Proceedings 163, Failure modes in composite materials with organic matrices and their consequences on design*, paper 1.

Potter, R. T. (1978) On the mechanism of tensile fracture in notched fibre reinforced plastics. *Proc. R. Soc. Lond.* A **361**, 325–41.

Rosen, B. W., Kulkarni, S. V. & McLaughlin, P. V. (1975) Failure and fatigue mechanisms in composite materials. *Inelastic Behaviour of Composite Materials*, ed. C. T. Herakovich, ch. 2. American Society of Mechanical Engineers, New York.

Shen, C.-H. & Springer, G. S. (1976) Moisture absorption and desorption of composite materials. *J. Comp. Mater.* **10**, 2–20.

Smith, T. R. & Owen, M. J. (1968) The progressive nature of fatigue damage in glass reinforced plastics. *Proceedings of the Sixth International Resins and Plastics Conference of the British Plastics Federation*, paper 27. British Plastics Federation, London.

Springer, G. S. (1977) Moisture content of composites under transient conditions. *J. Comp. Mater.* **11**, 107–21.

Whitney, J. M. & Nuismer, R. J. (1974) Stress fracture criteria for laminated composites containing stress concentrations. *J. Comp. Mater.* **8**, 253–65.

Main symbols

a, A area
a, b length of major and minor axes of ellipse
c crack length
C_{ij} compliance matrix
d diameter of fibre
E Young's modulus
G shear modulus
$G(\sigma)$ Weibull cumulative probability distribution function
K stress concentration factor, strength efficiency factor
l fibre length
M moment, moduli
P load, pressure
Q_{ij} reduced stiffness matrix
r radius of fibre
R half centre-to-centre spacing of fibres, stress ratio
s standard deviation, minimum distance between fibres
S span
S_{ij} compliance matrix
t thickness
V volume fraction
W work done per unit area in creating new surfaces, thermodynamic work of adhesion, weight fraction

α coefficient of thermal expansion
α, β angle
γ surface energy, shear strain
ϵ strain
η correction factor
θ angle
ν Poisson's ratio
ξ geometrical parameter depending on fibre shape and arrangement
ρ density, radius of curvature
σ stress
τ shear stress
ϕ winding angle

Subscripts

a angle-ply
A axial
c cross-ply
C compression
f fibre
F fracture
H hoop
i integer
l length
m matrix
mm modified matrix
N number fraction
o orientation
R polymer
s relating to shear strength of interface
T tension
v voids
W weight fraction
Y yield
$\parallel$ parallel to fibres
$\perp$ perpendicular to fibres
$\#$ shear, referred to the fibre and perpendicular directions
$\parallel\perp$ first index is direction of applied stress, second index is direction of contraction

Superscripts

$*$ value at fracture

The SI system of units

Base units

metre (m)	length
kilogram (kg)	mass
second (s)	time
ampere (A)	electric current
kelvin (K)	thermodynamic temperature
candela (cd)	luminous intensity

Some derived units

	Name of SI unit	*SI base unit*
frequency	hertz (Hz)	$1\ Hz = 1\ s^{-1}$
force	newton (N)	$1\ N = 1\ kg\ m\ s^{-2}$
work, energy	joule (J)	$1\ J = 1\ Nm$

Multiplication factors

	Prefix	*Symbol*
10^{12}	tera	T
10^{9}	giga	G
10^{6}	mega	M
10^{3}	kilo	k
10^{-3}	milli	m
10^{-6}	micro	μ
10^{-9}	nano	n
10^{-12}	pico	p
10^{-15}	femto	f
10^{-18}	atto	a

Some useful conversions

1 ångstrom $= 10^{-10}$ m = 100 pm or 0.1 nm
1 kgf = 9.8067 N
1 atm = 101.33 kN m^{-2}
1 torr = 133.32 N m^{-2}
1 dyne $= 10^{-5}$ N
1 dyne cm^{-2} = 1 mN m^{-2}
1 cal = 4.1868 J
1 erg $= 10^{-7}$ J = 0.1 μJ

1 inch	= 25.4 mm
1 lb	= 0.4536 kg
1 lbf	= 4.4482 N
1 lbf in^{-2}	= 6894.76 N m^{-2}
1 lb in^{-1}	= 0.1751 kN m^{-1}
1 eV	= 1.6×10^{-19} J = 0.16 aJ

Index